University Jubilees and University History Writing

Scientific and Learned Cultures and Their Institutions

Editor

M. Feingold (*California Institute of Technology*)

VOLUME 13

The titles published in this series are listed at *brill.com/slci*

University Jubilees and University History Writing

A Challenging Relationship

Edited by

Pieter Dhondt

BRILL

LEIDEN | BOSTON

Cover illustration: Festive procession on the occasion of the 500th anniversary jubilee of the University of Leipzig in 1909. Float with choir stalls and women carrying symbolic objects in front of it. Picture courtesy of the Stadtgeschichtliches Museum Leipzig.

Library of Congress Cataloging-in-Publication Data

University jubilees and university history writing : a challenging relationship / edited by Pieter Dhondt.
pages cm. -- (Scientific and learned cultures and their institutions, ISSN 2352-1325; volume 13)
Includes bibliographical references and index.
ISBN 978-90-04-21696-9 (hardback : acid-free paper) -- ISBN 978-90-04-26507-3 (e-book)
1. Universities and colleges--Europe--History. 2. Anniversaries--Europe--History. 3. Universities and colleges--Europe--Historiography. 4. Anniversaries--Europe--Historiography. 5. Historiography--Europe--History. 6. Europe--Intellectual life. 7. Europe--History, Local. I. Dhondt, Pieter.
LA621.U55 2015
378.4--dc23

2014034784

This publication has been typeset in the multilingual "Brill" typeface. With over 5,100 characters covering Latin, IPA, Greek, and Cyrillic, this typeface is especially suitable for use in the humanities.
For more information, please see www.brill.com/brill-typeface.

ISSN 2352-1325
ISSN 978-90-04-21696-9 (hardback)
ISSN 978-90-04-26507-3 (e-book)

Contents

List of Figures and Tables

Figures

Tables

List of Contributors

Thomas Brandt (1971)
is Associate Professor in History at the Norwegian University of Science and Technology (NTNU), Trondheim. His scholarly interests lie in the intersection between cultural history and the history of technology and science. Brandt's most recent book is a commissioned history of the Norwegian University of Science and Technology (2010).

Pieter Dhondt (1976)
is Senior Lecturer in general history at the University of Eastern Finland. He studied modern history at the KU Leuven and specialised in university history in Berlin and Edinburgh. After having obtained his doctoral degree at his home university, he has been working as postdoctoral researcher at the University of Helsinki and the *Université libre de Bruxelles*, and as lecturer in the history of education at Ghent University. His current research focuses on the history of university celebrations and on medical history, including the development of medical education at universities and colleges of higher education and the process of medicalisation in early childhood education. His publications deal with, among other themes, the intercultural transfer of university ideas within Europe in the nineteenth century and the history of academic mobility. Among his recent books: *Un double compromis. Enjeux et débats relatifs à l'enseignement universitaire en Belgique au XIXe siècle* (Gent: Academia Press 2011) and as editor *National, Nordic or European? Nineteenth-Century University Jubilees and Nordic Cooperation* (Leiden: Brill 2011).

Marek Ďurčanský (1973)
studied history and archival science at Charles University in Prague, where he also earned a doctorate in auxiliary historical sciences. He worked in the archives of the Academy of Sciences of the Czech Republic. Since 2005, he has led the archives of Charles University. His research focuses on the history of education and urban history. He is the editor of the anthology *Věda a Slovanství v 19. a 20. století* [*Science and Slavism in the 19th and 20th Century*] (Prague: Archiv AV ČR 2006), co-editor of the book *Korespondencja Tadeusza Kowalskiego z Janem Rypką i Bedřichem Hroznym* [*Correspondence of Tadeusz Kowalski with Jan Rypka and Bedřich Hrozný*] (Kraków: PAU 2007) and author of many scientific and popularising articles.

Jonas Flöter (1967)
secondary school teacher in history, social studies and economy in the Domgymnasium Naumburg. He studied history, art history and cultural sciences at the universities of Leipzig, Klagenfurt and Vienna, obtained his doctorate in 2001 and his habilitation in 2007. He is an outside lecturer (Priv.-Doz.) in philosophy and history of education at the University of Leipzig. Between 2006 and 2010 he was coordinator of the senate commission for the study of the history of the University of Leipzig. Current fields of research: Central European constitutional and political history, (social) history of education, third sector research and university history from the eighteenth to the twentieth century. Major publications: *Eliten-Bildung in Sachsen und Preußen. Die Fürsten- und Landesschulen Grimma, Meißen, Joachimsthal und Pforta (1868–1933)* (Köln: Böhlau Verlag 2009), *Leipziger Universitätsgeschichte(n). 600 Jahre Alma mater Lipsiensis* (Leipzig: Evangelische Verlagsanstalt 2009) and *Hugo Gaudig – Schule im Dienst der freien geistigen Arbeit. Darstellungen und Dokumente* (Bad Heilbrunn: Verlag Julius Klinkhardt 2012).

Jorunn Sem Fure (1967)
has studied social anthropology and history at the University of Bergen, where she also obtained her PhD. From 2002 to 2009 she has been working as senior researcher at the Forum for university history at the University of Oslo, where, among other engagements, she was responsible for two volumes of the history of the university, published in 2011: Volume 3, *1911–1940. Inn i forskningsalderen* and volume 4, *1940–1945 Universitetet i kamp*. Between 2009 and 2011 she was guest professor at the *Humboldt-Universität zu Berlin* and from 2012 she has been director of the Telemark Museum.

Trude Maurer (1955)
was educated at the universities of Tübingen and London. Dr. phil. Tübingen 1986; Habilitation Göttingen 1995. Researcher at the *Institut für Ost- und Südosteuropaforschung* (Regensburg) and lecturer in East European and Modern History at the University of Göttingen. Her publications include *Ostjuden in Deutschland 1918–1933* (Hamburg: Hans Christian 1986), *Die Entwicklung der jüdischen Minderheit in Deutschland. Neuere Forschungen und offene Fragen* (Tübingen: Niemeyer 1992) and *Hochschullehrer im Zarenreich. Ein Beitrag zur russischen Sozial- und Bildungsgeschichte* (Köln: Böhlau Verlag 1998). She co-authored *Jewish Daily Life in Germany, 1618–1945* (Oxford: University Press 2005) and edited collections of essays: on acculturation

processes in the Russian Empire (2000), on European universities during World War I (2006), on the complex relationship of universities and "their" cities in the Russian Empire (2009), and on the history of women's higher education (2010). Current research project: "'Russische' Doktorinnen deutscher Universitäten. Eine transnationale und transkulturelle Bildungsgeschichte."

Johan Östling (*1978*)
PhD in history from Lund University and Pro Futura Scientia Fellow at Swedish Collegium for Advanced Study (SCAS). His dissertation, *Nazismens sensmoral: Svenska erfarenheter i andra världskrigets efterdyning* [*The lessons of Nazism: Swedish experiences in the wake of the Second World War*] (Stockholm: Atlantis 2008), was awarded several prizes, including the Clio Prize and the Nils Klim Prize. As a postdoctoral researcher, he has been working at both the Research Policy Institute and the Department of History at Lund University. Östling's research interests combine intellectual, conceptual, political and cultural history, with a special emphasis on modern and contemporary European history. He is now working on a project on the transformation of the Humboldtian tradition and the idea of the university in twentieth-century Germany.

Emmanuelle Picard (*1967*)
is an associate professor of modern history at the *École normale supérieure de Lyon*. Her research interest focus on the history of French unversities, especially on the history of the French academic profession since the nineteenth century and the scientific assessment process. She is co-editor of *Histoire de l'éducation*. She also works on historical databases on European universities and took part in the foundation of Heloise (European Workshop on historical academic databases: heloise.hypotheses.org/). She is member of the scientific board of RESUP (*Réseau d'étude sur l'enseignement supérieur*). Her professional website is ens-lyon.academia.edu/EmmanuellePICARD.

Ana-Maria Stan (*1975*)
obtained her PhD in contemporary history in 2005 at the *Université Paris IV-Sorbonne* and the Babeş-Bolyai University of Cluj-Napoca, Romania. In her dissertation she investigated the political, cultural and economic relationships between France and Romania, in the troubled and difficult years of the Second World War, *La France de Vichy et la Roumanie* (*1940–1944*). *Collaboration et conflits* (Cluj: Académie Roumaine, Centre d'Études Transylvaines 2007). Currently, she works as a researcher at the Babeş-Bolyai University and manages the University Historical Museum. As a specialist of twentieth-century history, she

has published extensively on French-Romanian cultural and academic relationships, as well as on Romanian university life. In one of her latest projects, she joined an interdisciplinary postdoctoral research team that investigates the Romanian avant-garde from a political, socio-historical as well as literary perspective.

CHAPTER 1

Introduction

University History Writing: More than a History of Jubilees?

Pieter Dhondt

Researching and writing its own history has always been one of the tasks of the university. From the sixteenth century, rectors and ordinary professors have delivered speeches on the occasion of anniversary celebrations in which they presented their institution's glorious past. During the nineteenth century, the tradition of jubilee history reached its first peak. The historic speeches made on anniversary celebrations were gradually replaced by a thick and impressive series of commemorative volumes and *Libri Memorialis*, mostly written by history professors or sometimes by a special committee appointed by the rector. Of course, this kind of research concentrated on only one university, with little or no attention paid to other universities in the same country and even less so to those in other countries. In general, such homage offered a celebratory institutional history in which less savoury episodes were often disregarded. The tradition of jubilee history almost completely dominated university history until the Second World War and remained fairly popular even afterwards. However, in a series of two exploratory workshops funded by the Joint Committee for Nordic Research Councils for the Humanities and Social Sciences (NOS-HS), it was precisely these university jubilees which gave the initial impetus to an uncommon comparative approach.

The common title of both of these workshops was "University jubilees in Northern Europe: more than occasions to commemorate their own glorious past?" The first of them (in Helsinki, 12 and 13 March 2009) focussed on nineteenth-century university jubilees as the driving force for the development of Scandinavianist ideas and for increasing cultural and scientific cooperation between the Nordic countries. The workshop and the book[1] together was the realisation of the aim to encourage renewed cooperation between historians at different Northern European institutions. At the beginning of the twenty-first century, university jubilees still have the potential to bring the Nordic universities more closely together.

1 Pieter Dhondt (ed.), *National, Nordic or European? Nineteenth-Century University Jubilees and Nordic Cooperation* (History of Science and Medicine Library 25. Scientific and Learned Cultures and Their Institutions 4) (Leiden: Brill 2011).

 | DOI 10.1163/9789004265073_002

The aim of the second, historiographic workshop (in Copenhagen, 14 and 15 January 2010) was to study the challenges that face ongoing university history projects (many of them on the occasion of university jubilees), viz. to commemorate "their own glorious past" in a critical way, but without losing sight of relationships with other institutions at home and abroad. For several reasons, the publication of the results of this workshop was delayed and gradually this book developed a broader content, also by including some presentations from another international conference, in Ghent (16 and 17 March 2011) on the topic "Academic culture of remembrance: The combination of university history, jubilees and academic heritage". One of the main ambitions of this conference was to examine, from a historiographical perspective, the relationship between university history, the culture of remembrance in general, and university jubilees in particular. In consequence of this somewhat different approach, the original focus on Northern Europe has been forsaken.

However, the starting point in all the chapters in this book remains the traditionally close relationship between university history writing and university jubilees, as well as offering in the third part of the book some attempts to go beyond this customary practice. The relationship itself has seen many changes. When, from the end of the nineteenth century, the universities gradually developed into scientific research institutions, their culture of remembrance changed once again and in consequence, the related publications also took on another character. On the occasion of commemorations, books were edited with an emphasis on the (results of) scientific research from the past twenty-five, fifty or hundred years. Although these books were often rather dry and technical, simply listing the facts, this special kind of publication dealing with the history of the university had a specific aim: to illustrate the scientific output and to legitimise the indigenous institution in times of university expansion.

Frequently, by looking into its own history, the university as an institution led by traditions has tried to provide itself with "expertise concerning its future organization in a period of change".[2] By writing the history of its own institution, the university attempted to legitimise itself and to offer an answer to a certain crisis.[3] German university history writing, for instance, experienced a

2 Rüdiger vom Bruch, "Methoden und Schwerpunkte der neueren Universitätsgeschichtsforschung", in: Werner Bucholz (ed.), *Die Universität Greifswald und die deutsche Hochschullandschaft im 19. und 20. Jahrhundert* (Pallas Athene. Beiträge zur Universitäts- und Wissenschaftsgeschichte 10) (Stuttgart: Franz Steiner Verlag 2004): 11.

3 Stefan Gerber a.o., "Einleitung", in: Idem (eds.), *Traditionen – Brüche – Wandlungen. Die Universität Jena 1850–1995* (Köln: Böhlau Verlag 2009): 4.

flourishing period of jubilee histories in the 1960s and 1970s, not coincidentally in a period when the universities were confronted with all kinds of challenges: the massification of university education of course, but also the transition from a university controlled by a tenured faculty (*Ordinarienuniversität*) to a university where the main decisions were taken by the faculty board (*Gruppenuniversität*). The sense of crisis was sometimes even so intense that the question was raised on whether it was appropriate that the university ceremoniously celebrated the jubilee of its foundation. In several cases, the formal answer on the crisis – a jubilee history publication issued by the university authorities – was countered by a counter-festschrift written by students and/or a group of young scholars.[4] It is just one example of how jubilee history has been used as a weapon of propaganda.

Gradual Expansion of University History Writing in Three Directions

However, as dominant as jubilee history was and still is, gradually university history writing widened in three directions: geographically, thematically and chronologically the horizons were broadened. The initial impetus for a geographical widening was given by the German "patriarch of the discipline" Christoph Meiner, to use the words of the German historian Rainer A. Müller. In the first decade of the nineteenth century, Meiner published two volumes *Ueber die Verfassung und Verwaltung teutscher Universitäten* (Göttingen: J.F. Röwer 1801), and *Kurze Darstellung der Entwicklung der hohen Schulen des protestantischen Teutschlands, besonders der hohen Schule zu Göttingen* (Göttingen: J.F. Röwer 1808), as well as his magnum opus *Geschichte der Entstehung und Entwicklung der hohen Schulen unseres Erdtheiles*, 4 vols. (Göttingen: J.F. Röwer 1802–1805). In the preface of this latter work, Meiner emphasized that it was not his intention just to put together separate monographs on individual universities – as his predecessors in the seventeenth and eighteenth century had done – but instead he explained and defined his methodology of a 'structural history' of the university as a European phenomenon.[5]

4 Winfried Müller, "Erinnern an die Gründung. Universitätsjubiläen, Universitätsgeschichte und die Entstehung der Jubiläumskultur in der Frühen Neuzeit", *Berichte zur Wissenschaftsgeschichte* 21 (1998): 91–92.

5 Rainer A. Müller, "Genese, Methoden und Tendenzen der allgemeinen deutschen Universitätsgeschichte. Zur Entwicklung einer historischen Spezialdisziplin", *Mitteilungen der Österreichischen Gesellschaft für Wissenschaftsgeschichte* 20 (2000): 184.

It took until the last quarter of the nineteenth century before Meiner's initiative was followed by similar attempts in which the authors no longer limited themselves to the well-defined history of their own institution. In France, Louis Liard, general manager of the administration on higher education in the French republic, published *L'enseignement supérieur en France 1789–1889*, 2 vols. (Paris: Colin 1888–1894). And in Germany, as part of national propaganda for higher education, Friedrich Paulsen's *Geschichte des gelehrten Unterrichts auf den deutschen Schulen und Universitäten: Vom Ausgang des Mittelalters bis zur Gegenwart* (Leipzig: Veit & Comp. 1885) was one of a whole series of studies on a similar topic: the search for and the ode to the 'idea' or the 'essence' of the German university.[6] At the time of the Kaiserreich, both the German university and its self-contemplation through historiography were at the peak of their power and popularity.

From the 1980s, these national university histories began to be complemented with comparative studies from an international perspective, such as Fritz Ringer, *Education and Society in Modern Europe* (Bloomington: Indiana University Press 1979) and Konrad Jarausch (ed.), *The transformation of higher learning 1860–1930: expansion, diversification, social opening and professionalization in England, Germany, Russia and the United States* (Stuttgart: Klett 1983), to name just two of the earliest and most influential examples. Some specific topics in this context which developed from the 1990s are the study of the reception of foreign university systems (in practice, particularly the German model) and their relevance for internal reform processes,[7] and research on international student migrations. The massive student flows, mainly from Eastern Europe to Western European universities, clearly had repercussions on the development of education both in the country of origin and in the host

6 Müller, "Genese, Methoden und Tendenzen der allgemeinen deutschen Universitätsgeschichte" (2000): 188.

7 See, among many other studies: Willy Birkenmaier, *Das russische Heidelberg. Zur Geschichte der deutsch-russischen Beziehungen im 19. Jahrhundert* (Heidelberg: Verlag Das Wunderhorn 1995); Lothar Jordan and Bernd Kortländer (eds.), *Nationale Grenzen und internationaler Austausch. Studien zum Kultur- und Wissenschaftstransfer in Europa* (Communicatio 10) (Tübingen: Niemeyer 1995); Rainer Christophe Schwinges (ed.), *Humboldt International. Der Export des deutschen Universitätsmodells im 19. und 20. Jahrhundert* (Veröffentlichungen der Gesellschaft für Universitäts- und Wissenschaftsgeschichte 3) (Basel: Schwabe & Co. AG Verlag 2001) and Gabriele Lingelbach, *Klio macht Karriere. Die Institutionalisierung der Geschichtswissenschaft in Frankreich und den USA in der zweiten Hälfte des 19. Jahrhunderts* (Veröffentlichungen des Max-Planck-Instituts für Geschichte 181) (Göttingen: Vandenhoeck & Ruprecht 2003).

country.[8] Finally, the historic interest in the international relationship between universities is strengthened further by the current search for a unified European market of higher education.

Nevertheless, it remains difficult to break completely with old traditions, and in name international or national studies often focus on a country by country or university by university approach respectively. Also the fourth volume of the monumental *History of the University in Europe*, 4 vols. (Cambridge: University Press 1992–2011) (in itself, of course, a perfect illustration of the increasing geographical widening within university history) does not escape from this criticism.[9] This also shows the ever existing gap between what is on the one hand the still predominantly national approach within university history and, on the other hand, the much more internationally oriented history of science. Even though both sub-disciplines have been navigating their way towards each other in recent years, there is still a lack of studies in which they are really integrated. Whereas historians of science principally ask epistemic questions (and thereby go beyond the local or national context), questions of structural, social or political history (often from a national perspective) remain at the core of university history.[10]

Secondly, the geographical widening was gradually combined with a chronological one and, to a certain extent, even a chronological shift. In an article from 1978, the French historians Roger Chartier and Jacques Revel claim that before their time university history was in fact mainly medieval, institutional and intellectual history.[11] And even in countries where there existed no university during the Middle Ages, such as in the Northern Netherlands, almost by tradition the attention towards the late-medieval schools and universities stood at the centre, according to Willem Frijhoff.[12] His Dutch colleague, Willem Otterspeer, himself the author of a monumental history of the University of Leiden in the nineteenth century, has pointed out correctly that university

---

8 See, for instance, Hartmut Rüdiger Peter and Natalia Tikhonov (eds.), *Universitäten als Brücken in Europa. Studien zur Geschichte der studentischen Migration* (Frankfurt am Main: Peter Lang 2003).

9 Michael Shattock, "Review: A history of the university in Europe, Volume IV: Universities since 1945", *History of Education: Journal of the History of Education Society* 41 (2012), no. 3: 434.

10 Sylvia Paletschek, "Stand und Perspektiven der neueren Universitätsgeschichte", *Zeitschrift für Geschichte der Wissenschaften, Technik und Medizin* 19 (2011): 171.

11 Roger Chartier and Jacques Revel, "Université et société dans l'Europe moderne: Position des problèmes", *Revue d'Histoire moderne et contemporaine* 25 (1978): 353–374.

12 Willem Th.M. Frijhoff, "Hoezo universiteitsgeschiedenis? Indrukken en stellingen", *Ex Tempore. Historisch Tijdschrift KU Nijmegen* 15 (1996), no. 43: 13–22.

history writing since the 1980s has changed its main period of research, dealing preferably with the last two centuries.[13]

Finally, also the range of subjects expanded. The former almost exclusively institutional approach (celebratory or not) was supplemented with all kinds of other questions. The university was no longer considered an autonomous world detached from society, but instead the relationships between the university and the surrounding city (gown and town) and region became a special topic of study. The university was looked upon as a socio-economic space where knowledge and social skills were exchanged, traded or passed on. Historians became interested in the assaults on the ideological ideals of the university, changing in intensity and in nature, such as academic freedom, *Bildung* and the (acclaimed) unity of science. University history writing experienced the influence of a general switch towards the so-called new cultural history, paying attention to the history of the everyday things of academic life, to the self-conception of students and professors as separate social groups, the self-staging of universities on the occasion of celebrations, gender aspects, and the politicisation and radicalisation of the student body.[14]

Although from a historiographical point of view it took until after the Second World War before the university was really liberated from its ivory tower, not all of the above-mentioned subjects or approaches were completely new of course. Student history especially has a much longer (and stronger) historical background as a separate field of research, indeed of a somewhat ambivalent character. On the one hand it is one of the oldest independent branches within university history (e.g. already in the 1920s the still active German *Arbeitskreis der Studentenhistoriker* was founded, organising yearly conferences from 1924 onwards[15]), whilst on the other hand particularly this field has been dominated by popular, often sensationalist and distorted collections of anecdotes.[16] From the 1970s, the establishment of (semi-)professional societies of student history in several European countries contributed to the gradual emancipation of the field into a fully fledged academic research topic,

13 Willem Otterspeer, *Het achtste memo. Over begin en einde, over universiteit en universiteitsgeschiedenis* (Leiden: Rijksuniversiteit 1997).

14 Matthias Asche and Stefan Gerber, "Neuzeitliche Universitätsgeschichte in Deutschland. Entwicklungslinien und Forschungsfelder", *Archiv für Kulturgeschichte* 90 (2008), no. 1: 199–201, including a number of specific titles by way of example.

15 See www.burschenschaftsgeschichte.de/studentenhistoriker/arbeitskreis.htm (date accessed 11/06/2014).

16 Asche und Gerber, "Neuzeitliche Universitätsgeschichte in Deutschland" (2008): 196–198.

resulting in conferences such as "*Universitas scholarium.* The social and cultural history of the European student from the Middle Ages to the Present" (Cologne, 24–25 October 2013) or the ongoing book project launched by the International Commission on the History of Universities "Student Revolt, City and Society – From the Middle Ages until Today". However, just as in the case of university history writing in general, the balance between an engaged, celebratory approach and a more distant, academically critical one is still sometimes precarious.[17]

Spurred on by this geographical, chronological and thematic widening, university history gradually developed into a separate discipline, morphing into a sub-genre of the history of science. Independent research institutes[18] and specialised societies[19] were founded, often publishing their own journals.[20]

---

17 A series of jubilee books on Finnish student societies, published in the 1990s and early 2000s, can be considered successful examples of the search for such a balance, for instance, all the authors being or having been actively engaged in their object of study, yet at the same time having obtained an academic degree in history. E.g. Jari Hanski, *Polin suojiin me saavumme taas. Teknillisen Korkeakoulun Ylioppilaskunta 125 vuotta* [*125 years of the Student Union of the Helsinki University of Technology*] (Espoo: Teknillisen Korkeakoulun Ylioppilaskunta 1997); Kimmo Ketonen, *Sun kasvois eessä, suomenmaa. Turun Yliopiston Ylioppilaskunta 1922–1944* [*The student union of the University of Turku 1922–1944*] (Turku: Turun yliopiston ylioppilaskunta 1997); Reijo Valto, *Mitä teekkarit eilen... 1969–1999. Lappeenrannan teknillisen korkeakoulun ylioppilaskunta 1969–1999* [*The student union of the Lappeenranta University of Technology 1969–1999*] (Lappeenranta: Lappeenrannan teknillisen korkeakoulun ylioppilaskunta 1999); Kimmo Ketonen, *Ylioppilaat omalla asialla: Turun yliopiston ylioppilaskunta 1945–1997* [*The student union of the University of Turku 1945–1997*] (Turku: Turun yliopiston ylioppilaskunta 2001) and Marko Lamberg, *Nuoruus ja toivo. Jyväskylän yliopiston ylioppilaskunta 1934–2003* [*The student union of the University of Jyväskylä 1934–2003*] (Jyväskylän yliopiston julkaisusarja 70) (Jyväskylä: Kampus 2004).

18 E.g. *Abteilung Wissenschafts- und Universitätsgeschichte* in Munich (1999–2003), The International Centre for the History of Universities and Science in Bologna (www.cis.unibo.it/), *Centre Interuniversitario per la storia delle università italiane* (www.cisui.unibo.it/); all websites accessed 11/06/2014.

19 E.g. International Commission for the History of Universities (from 1960, www.cihu-ichu.org/), *Gesellschaft für Universitäts- und Wissenschaftsgeschichte* (from 1995, guw-online.net/), Gewina. the Belgian-Dutch Society for the History of Science and Universities (from 1995, www.gewina.nl/), Research network on the history of university and higher education in Finland (from 2007, www.helsinki.fi/historia/ylhist/); all websites accessed 11/06/2014.

20 E.g. CIAN-Journal of the History of Universities (from 1998, e-revistas.uc3m.es/index.php/CIAN), *Jahrbuch für Universitätsgeschichte* (from 2000, www.steiner-verlag.de/JbUG/),

Many of these initiatives take care to assemble bibliographies on university history in a national or regional context. Since 1981, the journal *History of Universities* has attempted to accumulate these bibliographic notices as a kind of coordinating initiative, thoug with varying success. Jubilees no longer only result in the publication of commemorative books, but also provide the basis for the establishment of extraordinary chairs, such as at Dutch universities,[21] or the foundation of true research centres such as the Forum for University History in Oslo[22] or the *Senatskommission zur Erforschung der Leipziger Universitäts- und Wissenschaftsgeschichte* in Leipzig, yet the fate of these initiatives in the long term is not always assured (see the chapter of Jonas Flöter). Special series were started, some of which became clearly authoritative in the field, such as *Contubernium. Tübinger Beiträge zur Universitäts- und Wissenschaftsgeschichte*, which includes already more than 75 volumes, many of which are not geographically restricted to Tübingen. Of course, the connection with the history of science and the predominance of German-speaking scholars in this regard are far from accidental.

Through such societies, journals and conferences, the geographical widening with regard to the content of university history writing is increasingly combined with a practical geographical widening. University historians from different (mainly European) countries meet together resulting sometimes in huge projects. A good example is the network "Heloise, European Workshop on Historical Academic Databases," established in 2012, yet with a background going back to the late 1960s.[23] The aim of the network is to coordinate and

Annali di Storia delle Università italiane (from 1997, www.cisui.unibo.it/frame_annali.htm), *Studium* (from 2008, www.gewina-studium.nl/); all websites accessed 11/06/2014.

21 In recent years most of the Dutch universities have appointed special university historians: Peter Jan Knegtmans at the University of Amsterdam (since 1995), Willem Otterspeer in Leiden (since 1997), Leen Dorsman in Utrecht (since 2001), Jan Brabers at the Radboud University Nijmegen (since 2002), Klaas van Berkel at the Rudolf Agricola chair in Groningen (since 2008) and Ab Flipse at the VU University Amsterdam (since 2013). According to Marc Wingens, former university historian at the VU University Amsterdam, the origin of the universities' interest in their own history can be found in the conclusion that they increasingly resembled each other in consequence of the developments during the previous decades. Therefore, the university administrations were searching for ways to distinguish their own institution from those of their colleagues. The own past indeed offered a nice possibility to create a distinct profile of the own institution. Marc Wingens, "Stand van zaken van de projecten geschiedschrijving aan de Nederlandse universiteiten, jan. 2001", *Nieuwsbrief Universiteitsgeschiedenis* 7 (2001), no. 1: 7–17.

22 See www.hf.uio.no/iakh/english/research/projects/ffu/ (date accessed 11/06/2014).

23 Willem Frijhoff, "Avant Héloïse: L'aventure de Fasti" (13/09/2012), heloise.hypotheses.org/75 (date accessed 11/06/2014).

streamline separate initiatives to breathe new life (on the internet) into the traditional nineteenth-century *Libri Memorialis* in the form of inventories and prosopographic databases of professors[24] and students (see the chapter of Emmanuelle Picard).

Thus, current university history writing is aiming for much more than offering just a set reproduction of a fragment of the past. In addition, it holds up a mirror to the university when reflecting on its own social and scientific role in the future. Nevertheless, an important limitation remains the almost exclusive focus on the history of universities, neglecting the relationships with other institutions of higher education. Moreover, the history of education is also not dealing with this level of education, so the history of colleges of advanced education is almost completely left out in the cold. And what is more, in general both historic sub-disciplines are still much too separated from each other, even though the borders between secondary and tertiary education are not always as unequivocal, and methodologically university history and history of education could clearly benefit from each other (see the chapter of Pieter Dhondt). A close cooperation between historians, who study these different levels of education, could produce some worthwhile results.

Continuing Tradition of Jubilee History

And yet, despite all these impulses for a renewal of the field, jubilee history remains an exceptionally crucial genre within university history writing. The tradition of jubilee history is still very much alive: focussed on one celebrating institution, written by members of its own academic community and, especially in times of crisis, presented as something one can learn from, at least implicitly. The challenge for all historians commemorating the past of their indigenous institution is to take into account the above-mentioned scientific developments, but at the same time to have an eye for the component of visibility, appearance and communication towards the broader public, strived for by the university as a living institution.

24 E.g. Amsterdam (www.albumacademicum.uva.nl/), Gent (www.ugentmemorialis.be/), Halle (www.catalogus-professorum-halensis.de/), Helmstedt (uni-helmstedt.hab.de/), Kiel (still in the making, www.uni-kiel.de/aktuell/pm/2012/2012-099-professorenkatalog.shtml), Leipzig (www.uni-leipzig.de/unigeschichte/professorenkatalog/), Rostock (cpr.uni-rostock.de/), Utrecht (profs.library.uu.nl/index.php); all websites accessed 11/06/2014. See also Gregor Horstkemper and Alessandra Sorbello Staub, "Professorenkataloge, Vorlesungsverzeichnisse, Hochschulreden. Materialien zur Universitätsgeschichte im Internet", *Geschichte in Wissenschaft und Unterricht* 61 (2010), no. 2: 130–131.

In that way, university history still has to emancipate itself, at least to a certain extent, from the field from which it originates and to which it, to this day, relates itself both problematically and productively: the culture of university jubilees. On the one hand university history certainly can make use of the occasion of jubilees to obtain some extra funding, but on the other hand the jubilee can not misuse university history to rewrite the institution's history.[25] The quote of an American author in the memorandum *Hochschule in der Demokratie* from 1965, that "a university is an institution that studies systematically [meaning 'critically' in this context] almost all that exists under the sun – except for itself,"[26] nowadays no longer holds true. However, it is still an important disadvantage that the current tradition of writers of jubilee histories almost always focus on their indigenous institution, which makes it sometimes difficult to maintain enough distance and to deal with current episodes, also because this includes discussing the experiences and achievements of one's own academic teachers and colleagues and having to put aside one's own subjective impressions.[27]

The outcome of the current tradition of jubilee history writing are two kinds of publication that can be placed opposite to one another as follows, indeed in a somewhat caricatured way: on the one hand extremely detailed factual accounts of the university as an institution, combined with some developments in separate scientific disciplines, written collectively by a number of authors and published in several volumes; and on the other hand abridged 'coffee table synthesis', written by only one author, with a number of pictures and generally ignoring the unsavoury episodes in the institution's past. Different arguments can be put forward for moderate versions of both types of publications existing alongside each other, and even if a consensus could be found, it would probably lie somewhere in the middle. Nevertheless, independent of the final choice, a general agreement among university historians is growing that both kinds of stories are absolutely in need of a clear master narrative. The Oxford professor Laurence Brockliss endeavours to achieve this in his forthcoming history of his own institution, to be published in 2014 in one

25 "Jubiläum ohne zu jubilieren" was the title of a lecture from Eva-Maria Felschow, university archivist in Giessen, which she gave in Jena in 2002 with regard to the upcoming university jubilee in Giessen in 2007. Cited after Gerber a.o., "Einleitung" (2009): 3. See also Notker Hammerstein, "Jubiläumsschriften und Alltagsarbeit. Tendenzen Bildungsgeschichtlicher Literatur", *Historische Zeitschrift* 236 (1983): 601–633.

26 Wolfgang Nitsch a.o., *Hochschule in der Demokratie. Kritische Beiträge zur Erbschaft und Reform der deutschen Universität* (Berlin: Luchterhand 1965): 1.

27 Müller, "Erinnern an die Gründung" (1998): 94.

volume, with as a leitmotiv a story of continuity and discontinuity.[28] Sverker Sörlin and Nina Wormbs compose their story about the Royal Institute of Technology in Stockholm around that which the institution itself claims to be offering: innovation and impact on the society.[29]

But also in multi-volume histories, the reader should be kept enthralled through the use of one (or a few) master narrative(s) common for all the volumes. The 9-volume history of the University of Oslo is a good example in this respect.[30] The project was launched shortly after the university went through an institutional crisis and, certainly for the administration, one of the central questions was how the disciplinary diversity came into existence and how the university itself could be maintained? How can the university organise itself so that it can act as an entity, as a unity in a changing world? In answer to this, Edgeir Benum (initial manager of the project) expressed the integrative ambitions of the upcoming university history in question form as follows: "What has held parts of the university together, what provided a sense of belonging to more than one's own little mini-state?"[31] How the university has been coping with the problem of diversity and unity, on the level of disciplines, but also on the level of student and staff (e.g. the increasing classification in different groups of personnel), and on the level of the competing tasks of the university?

A year later, Robert Marc Friedman explicitly reconfirmed these integrative ambitions:

> We will resist the practice of farming out portions of the history to the alleged disciplinary 'specialists' for each aspect of the university's history. We do not want somebody to write a narrow institutional history to which is added in separate chapters the students, the women, the natural sciences or even each discipline in turn, and so forth. There is a biography

---

28 Pieter Dhondt and Christophe Verbruggen, "Academic culture of remembrance. The combination of university history, jubilees and academic heritage", *Studium. Tijdschrift voor Wetenschaps- en Universiteitsgeschiedenis/Revue d'Histoire des Sciences et des Universités* 5 (2012), no. 3: 140.

29 Sverker Sörlin and Nina Wormbs, "KTH and the Formation of Swedish Innovation Systems, 1920 to 1980", presentation at the conference "Rethinking Modern University History" (Oslo 29/10/2011).

30 John Peter Collett (ed.), *Universitetet i Oslo 1811–2011*, 9 vols. (Oslo: Unipub forlag 2011).

31 Benum, Edgair, *On the challenge of writing a university history: The University of Oslo* (Research & Occasional Paper Series: CSHE.5.99) (Berkeley: University of California, Center for Studies in Higher Education 1999): 9, cshe.berkeley.edu/challenge-writing-university-history-university-oslo/(date accessed 11/06/2014).

> of a university waiting to be uncovered; not merely the sum of biographies of its professors.[32]

The (besides impressively large) group of Norwegian scholars was probably not the first who aimed to write an integrated history of their own institution, but they were certainly among the first to express this so explicitly. The one volume history of the University of Jena of 2009 was also intended as an integrated biography, reviving in this way the identity of the university as a concept of unity and universality.[33] According to a well-known university historian from Freiburg, Sylvia Paletschek, the Jena colleagues were equally successful in doing so.[34]

Especially with regard to the development of the disciplines, this integrated approach often proves to be a great challenge. Indeed, nobody can deny the increasing specialization and fragmentation of scientific research, resulting in the so-called 'multiversity',[35] and an approach per faculty or per discipline is therefore clearly justifiable. In result, several university history committees have opted for a more fragmented kind of publication, not so much because this approach is easier, but rather as a well-considered approach in the current climate of multiversity. In a somewhat traditional way, one of the volumes published on the occasion of the bicentenary of the *Humboldt-Universität zu Berlin* in 2010 is composed as a separate history of the different disciplines.[36] The forum *Zeitgeschichte der Universität Wien*, established in 2011 to prepare the 650th anniversary of the university of Vienna in 2015, flirts with the idea of combining in its 'faculty volume' a historical overview (focussed on how the disciplinary differentiation took place and why it occurred as it did), with the creation of a historic source by asking the key figures in the different scientific fields to reflect on the current position of their field and in which direction it is to develop in the future.[37]

---

32 Robert Marc Friedman, *Integration and Visibility: Historiographic Challenges to University History* (Oslo: University of Oslo 2000): 44.

33 Gerber a.o., "Einleitung" (2009): 15.

34 Paletschek, "Stand und Perspektiven der neueren Universitätsgeschichte" (2011): 183.

35 Clark Kerr, *The Uses of the University* (Harvard: University Press 1963).

36 Daniel Hechler and Peer Pasternack, "Best Practice vs. Worst Case? How East German Universities Deal with their Contemporary History: The Humboldt University of Berlin and the University of Jena", *Berichte zur Wissenschaftsgeschichte* 34 (2012): 329–345. See also Daniel Hechler and Peer Pasternack, *Traditionsbildung, Forschung und Arbeit am Image: Die ostdeutschen Hochschulen im Umgang mit ihrer Zeitgeschichte* (Leipzig: AVA-Akademische Verlagsanstalt 2013).

37 See www.forum-zeitgeschichte.univie.ac.at/universitaet/forum-zeitgeschichte/ (date accessed 11/06/2014).

The commission preparing the bicentenary of Ghent University in 2017 has a similar goal with UGentMemorie, the virtual memory of Ghent University (www.UGentMemorie.be/, date accessed 11/06/2014): combining university history writing with the creation of historic source material. By drawing from insights in public history, academic heritage and digital history, they explore ways of filling in the gap between historians and communicators, the demands of science and the expectations of the public. In their view, public and digital history are interesting ways of communicating research on university history and heritage on different levels on the one hand, and to integrate in a sensible way the academic culture of remembrance in a historical project on the other. In terms of output, this means that the classic publication of the complete history of the university at least has to be accompanied by other initiatives, such as smaller occasional publications, exhibitions or heritage walks, digital output and social media strategies.[38]

Also Klaas van Berkel promises an original approach in his history of the University of Groningen (to be published in 2014) by combining his master narrative (i.e. the university as an academic living community) with the use of an anthropological methodology. The central questions are "how the academic living community (of teachers, students, administrators), enables the university to endure one crisis after the other?" and "what makes a university into a university, as a living community?" In order to write such a history, the historian can not limit himself to the usual conventions of institutional, intellectual and social history. Therefore, Van Berkel takes an essential additional method from anthropology: every so often through a meticulous description of the everyday customs and habits, he will attempt to find out why the members of the academic community behave as they do. Regularly he will make an appeal to the *pétite histoire* of the university, the small anecdotes and the dodgy stories, but by bringing out the contrast with the 'official' story, he will be able to develop an insight into how the academic community really functions.[39]

By accepting the assignment of writing the history of his own institution on the occasion of its quartercentenary, Van Berkel confirms a trend that he has regretted on other occasions: the limitation of university history to

38 Fien Danniau, Ruben Mantels and Christophe Verbruggen, "Towards a Renewed University History: UGentMemorie and the Merits of Public History, Academic Heritage and Digital History in Commemorating the University", *Studium. Tijdschrift voor Wetenschaps- en Universiteitsgeschiedenis/Revue d'Histoire des Sciences et des Universités* 5 (2012), no. 3: 179–192.

39 Klaas van Berkel, *Academisch leven. Over geschiedenis, karakter en veerkracht van de Nederlandse universiteit* (Amsterdam: Bert Bakker 2009).

publications produced on the occasion of jubilees.[40] However, the ambition to enlarge the field with projects of a comparative and transnational approach is clearly more easily discussed than realised.[41] The series of exploratory workshops from which this book originates have certainly contributed in this attempt, through the resulting publications (comprising an explicitly comparative and transnational approach in the first volume, with a somewhat more implicit method in the second volume, as will be explained below) and through a general increase in the cooperation between university historians from all the Nordic countries. In that way, the following conferences (and possible publications) can be considered an indirect outcome of the workshops, or at least events confirming the taken path: "Elitist institutions in egalitarian societies? Visions and realities of Nordic universities" (10–12 November 2010, Berlin), "The Humboldtian tradition – Origin and legacy" (24–25 November 2010, Uppsala)[42] and "The transformation of the Nordic universities in the 1960s and 1970s – A re-examination of the 'student revolution'"; special session at the 27th Congress of Nordic Historians (11–14 August 2011, Tromsø).

University Jubilees and University History Writing

The first part of the book focuses on the traditional relationship between university jubilees and university history writing. The starting point in each case is the history of the jubilee itself, studying its general message and its political interpretation; political in a double sense: on the one hand the course and the content of the jubilees were highly dependent on political developments, but on the other hand the jubilees themselves often also disseminated a political message, e.g. through the conferment of honorary degrees. However, all the authors pay special attention to the way in which this general objective of the jubilee (being of a political, cultural and/or scientific kind) has been supported through specific jubilee publications; how the authors of these publications have been mobilised for the purpose of the common goal of the university

40 Dhondt and Verbruggen, "Academic culture of remembrance" (2012): 140.

41 E.g. Pernilla Jonsson and Silke Neunsinger, "Comparison and transfer – A fruitful approach to national history?", *Scandinavian Journal of History* 32 (2007), no. 3: 258–280 and Paul Mattingly, Konrad Jarausch, John Craig, Joseph Kett and James Turner, "Universities in Europe: North American Perspectives on European Historiography", *History of Education. Journal of the History of Education Society* 37 (2008), no. 3: 469–490.

42 Thomas Karlsohn, Peter Josephson and Johan Östling (eds.), *The Humboldtian Tradition: Origins and Legacies* (Scientific and Learned Cultures and Their Institutions 12) (Leiden: Brill 2014).

authorities or political administration in organising or controlling the jubilee; and how jubilee history has been used as a weapon of propaganda.

Marek Ďurčanský and Pieter Dhondt show in the first case study how the outcome of the jubilees of Charles University in Prague in 1848 and 1948, including the historical publications, works of art, and building alterations, have been influenced by unexpected political events. At the centenary celebration of the Royal Frederik University in Kristiania (Oslo) in 1911, the topic of Jorunn Sem Fure's chapter, political as well as scientific issues were at stake. Especially the latter struggle was fought out through the jubilee publications, among other means: Bredo Morgenstierne's plea for the university as an institution for the education of the people in his extensive history of the university, against a defence of the university as a research institution by the then rector, Waldemar Christopher Brøgger, in a pamphlet with an obvious historical slant to it. In the third chapter Trude Maurer compares the celebration of patriotic anniversaries in Russian and German universities in 1912/13. Indeed, although she is not dealing with university jubilees as such, again it becomes clear to what extent history writing was used to enforce the general message of the celebrations. The last chapter of this part focuses on the academic ceremonies and celebrations at the Romanian University of Cluj between 1919 and 2009. Ana-Maria Stan investigates the diverse forms of festivities that shaped a specific institutional identity for this university, outlining continuities and fractures, often determined by political regimes. Each academic jubilee provided a different context of reflection on the subject, giving birth to many types of publications or in-house events.

The last celebration under review in Stan's chapter took place in 2009, and in that way she makes an excellent link to the second part of the book, in which two recent experiences of writing the history of the own institution on the occasion of a jubilee are discussed. The Norwegian University of Science and Technology, celebrating its centenary in 2010, functions in this regard as a typical example of a merged institution struggling over what parts of the history ought to be treated and which period to be focussed on, as is shown by Thomas Brandt. The experiences in Leipzig, on the occasion of its 600th anniversary jubilee in 2009, firstly prove how crucial it is to ensure the continuity of the investments in university history writing on a financial and personnel level also after the jubilee, of course on a lower profile, but the efforts of both the temporarily appointed collaborators in the project and the university archivists[43] should not be limited to a flash in the pan. Secondly, the article by Jonas

43 Jens Blecher and Marek Ďurčanský, "Universitätsjubiläen und Universitätsarchive. Die Jahrhundertfeiern an den Hochschulen in Prag und Leipzig als Chance für die Universitätsarchive", *Jahrbuch für Universitätsgeschichte* 14 (2011): 229–234.

Flöter makes clear the extent to which some kind of glorious rhetoric seems to be embedded within this type of jubilee history writing. In his first sentence, for instance, Flöter praises the university for its long historical background. Moreover, the Nobel Prize winners can not be omitted from the extremely short historical overview of the university's development, which is concluded with a positive outlook upon the future: "In result of these measures, the University of Leipzig has developed a new, strong performance profile since the mid 1990s – to become a modern, cosmopolitan and future oriented university." This also explains the only implicitly comparative tone of this book, at least with regards to its first two parts. University history writing as part of the jubilee and on the occasion of a jubilee is by nature limited to one institution only, without a clear comparative or transnational approach.

However, the third part of the book aims to go beyond the jubilee and to offer three other ways of writing university history. By making a clear synthesis of the way in which German universities have been dealing with the Humboldtian tradition between 1810 and 1945, Johan Östling shows convincingly that German historiography in particular is absolutely not limited to jubilee publications. On the other hand, the Humboldtian tradition is inextricably bound up with the jubilee tradition, if only because the name of Humboldt has been invoked at a large number of anniversary celebrations since the beginning of the twentieth century all over the Western world. In addition, the aim of the article is to give a general, overarching historical background to the development of an academic self-understanding, which not least has been manifested at various university jubilees.

Emmanuelle Picard links up to Brandt's chapter in the sense that French universities are also still struggling over proper jubilee celebrations due to several ruptures since the French Revolution and up to the events of 1968, resulting in the breakup of existing universities into multiple entities without their own past. By way of alternative Picard suggests not writing a traditional jubilee history of a university, but instead developing a prosopographic project on the academic teaching corps in order to foster a genuine social history of the academic profession and to fully understand French academia as an institution. In the closing chapter, Dhondt also focuses on teaching as one of the central tasks of the university. The main goal of the chapter is to explain what university history as a sub-discipline could gain by connecting itself more closely to the methodologies and concepts used in the field of history of education.

This specific approach is largely inspired by my experience as lecturer in the history of education at Ghent University between 2009 and 2012, and more particularly through my close contacts with Marc Depaepe, by whom I actually started my academic career 14 years ago. Marc surely has sometimes regretted my choice of focussing on university history instead of history of education, but he never did blame me for it and always kept on supporting me. However, my thanks not only go to him, but first of all to Robert Marc Friedman, Guðmundur Hálfdánarson and Ejvind Slottved, my co-applicants at the Joint Committee for Nordic Research Councils for the Humanities and Social Sciences. Of course, the NOS-HS itself cannot be missing from these words of gratitude for its generous financial support of both exploratory workshops and the publications resulting out of them. I would like to thank the contributors in this volume for their work, but also equally for their patience. Indeed, the project lasted much longer than all of us expected, and the way it reached its completion has not always been easy. One of the causes of this was my continual change from one university to the other, a typical example of a postdoctoral scholar wandering around in the competitive European academic market. On this odyssey, I could always fall back on the support of a group of colleagues from different countries that have become increasingly like friends: Hilde de Ridder-Symoens in the first place, Laura Kolbe, Leen Dorsman, Jo Tollebeek and Robert D. Anderson. My warmest thanks for your continuous trust and support. Finally, I would not have been able to have sustained such an array of projects and changes in working environment without the support of my wife. Thank you for this, and for your practical assistance.

Pieter Dhondt
Joensuu, 11 June 2014

PART 1

University History Writing as Part of the Jubilee

CHAPTER 2

Two Great Anniversaries, Two Lost Opportunities

Charles University in Prague, 1848 and 1948

Marek Ďurčanský and Pieter Dhondt

The celebrations of both centenaries of Prague University in 1848 and 1948 (500 and 600 years since its foundation) represent a specific kind of interrupted university jubilee. In both cases, the organizers began their work sufficiently in advance, but the result was fundamentally altered by unexpected political events. The absence (in 1848), or marked ideological distortion (in 1948) of the central festive ceremony had a negative impact, especially on the anticipated international reaction to the jubilee of the oldest studium generale *in Central Europe. Planned programmes of both celebrations extensively drew on a tradition of commemorating Emperor Charles IV, the founder of the university. Plans also included most of the elements typical for university jubilees of the nineteenth and twentieth century: awards of honorary doctorates, jubilee publications, new works of art connected with the university, and alterations to important university buildings. In both cases, a crucial role was assigned to the* Karolinum, *the oldest building of the university in the very centre of Prague, which had been used for academic purposes since medieval times. In the end, due to the political circumstances, it was the publications, works of art, and building alterations that formed a lasting contribution to the university tradition. In 1848, this is most clearly the case of the majestic monument to Charles IV close to the Charles Bridge, which was the most expensive project of the quinquecentennial celebrations. A hundred years later, the jubilee enabled the university to gain the financial means for a structural survey and well-considered reconstruction of the* Karolinum.

In recent decades, festivities and rituals connected with them are among the most studied subjects in cultural history. Obviously, academic festivities are part of this topic and it is not surprising that special attention has also been paid to anniversaries of various universities and the celebrations related to them.[1] Winfried Müller has shown that in the early modern era, universities

1 Jens Blecher and Gerard Wiemers (eds.), *Universitäten und Jubiläen. Vom Nutzen historischer Archive* (Veröffentlichungen des Universitätsarchivs Leipzig 4) (Leipzig: Universitätsverlag 2004); Pieter Dhondt (ed.), *National, Nordic or European? Nineteenth-Century University Jubilees and Nordic Cooperation* (History of Science and Medicine Library 25. Scientific and Learned Cultures and their Institutions 4) (Leiden: Brill 2011).

 | DOI 10.1163/9789004265073_003

played an important role in transferring the originally ecclesiastic setting of jubilees to a lay environment.[2] The outer form of university anniversaries and their celebrations began to receive a stable, fixed appearance in the eighteenth and especially in the nineteenth century, when this kind of event experienced a remarkable boom.[3] From that time onwards, one can hardly find university anniversaries of any importance that were not accompanied by jubilee publications, the unveiling of a memorial, reconstructions of university buildings, participation of foreign guests in festive ceremonies, awards of honorary doctorates, and the like. University anniversaries attracted the attention of the public and that is also the reason why this potential has often been used for political purposes. Such anniversaries therefore often also functioned as celebrations for the ruling dynasty, nation, political party, or regime and conversely, a change in the political situation could radically disrupt the concept and preparations of university jubilees or even thwart them altogether.[4]

It was precisely this that happened to both of the large celebrations of round anniversaries of Prague University that are the subject of this study. Due to turbulent political events that significantly limited the international impact of the jubilees in particular, the celebrations could not proceed according to long-prepared plans. Even so, they were important for the university itself because some of the projects related to them produced results, which became a lasting part of the academic tradition in Prague.

Karolinum – a Building as a Symbol of Academic Tradition

For centuries, it has been customary for successful graduates in Prague to celebrate the completion of their studies by participating in a pompous graduation ceremony. On their way to the great hall of the university, graduates pass over a small but prominent metal plate built into the brick floor. The engraved

2 Winfried Müller, "Erinnern an die Gründung. Universitätsjubiläen, Universitätsgeschichte und die Entstehung der Jubiläumskultur in der frühen Neuzeit", *Berichte zur Wissenschaftsgeschichte* 21 (1998): 79–102.

3 Thomas P. Becker, "Jubiläen als Orte universitärer Selbstdarstellung. Entwicklungslinien des Universitätsjubiläums von der Reformationszeit bis zur Weimarer Republik", in: Rainer Christoph Schwinges (ed.), *Universität im öffentlichen Raum* (Veröffentlichungen der Gesellschaft für Universitäts- und Wissenschaftsgeschichte 10) (Basel: Schwabe Verlag 2008): 77–107.

4 For more on university anniversaries that were to an important degree affected by political and other external circumstances, see Sylvia Paletschek, "Festkultur und Selbstinszenierung deutscher Universitäten", in: Ilka Thom and Kirsten Weinig (eds.), *Mittendrin. Eine Universität macht Geschichte* (Berlin: Akademie Verlag 2010): 92–95.

dates of 1348 and 1948 are meant to recall not only Prague's long academic tradition, but also the most recent centenary celebration. These graduation ceremonies take place in the *Karolinum*, the oldest building of Prague University, located in the very centre of Prague's Old Town (see figure 2.1).[5]

Karolinum is doubtlessly an important place of memory[6] within Central Europe as a whole. Originally a burgher palace, it was acquired by the university as early as the fourteenth century and it quickly became the heart of its life, functioning as the setting of academic festivities from medieval times. Only between 1654 and 1773, when the faculties of philosophy and theology of Prague University were entrusted to the Jesuit order, did the Jesuit college of Saint Clement (the *Klementinum*), less than one kilometre away, compete with the *Karolinum*. However, after the dissolution of the Jesuit order, *Karolinum* again became the seat of the rectorate of the united university, which until 1918 was named not only after its founder, Charles IV, but also after the Habsburg Emperor Ferdinand III (thus bearing the name Charles-Ferdinand University).

FIGURE 2.1 *The* Karolinum *around 1900.*
PICTURE COURTESY OF THE INSTITUTE OF THE HISTORY OF CHARLES UNIVERSITY AND THE ARCHIVES OF CHARLES UNIVERSITY

5 See Josef Petráň, *Karolinum* (Prague: Karolinum 2010).

6 Pierre Nora, *Les lieux de mémoire*, 7 vols. (Paris: Gallimard 1984–1992).

The great hall of the *Karolinum* is nowadays not only a place where regular academic rituals take place (such as matriculations, graduations and inaugurations of academic officials). It also hosts ceremonies of national and international importance, such as the reception of official foreign state visits and the inauguration of professors of Czech universities by the president of the republic.

It is thus not surprising that the *Karolinum* always played a central role in plans for jubilee celebrations of Prague University. And vice versa, jubilees largely influenced the external appearance of the building, which holds especially true for the 1948 anniversary. It needs to be emphasized that the organizers designed both of the two most recent centenary anniversaries of Prague University as events with a nationwide and even international impact. The university had seen times of glory and decline, but its importance almost always reached beyond the borders of the national state and the jubilee celebrations were supposed to reflect this fact. And it was precisely the *Karolinum*, the name of the building referring to Charles IV, which functioned as the main visual symbol of the international status of Prague University. As will become clear, this traditional reference to Charles IV was one of the key elements in both centenary celebrations.

Preparations for the 1848 Anniversary

On the eve of the university's quinquecentennial in 1848, Prague University was a state institution subjected to strict bureaucratic and police supervision.[7] Its purpose was the training of state employees, clerics, and mainly lawyers and physicians.[8] Whereas as recently as in 1784, students of theology still constituted almost half of the total number of 1.192 students, in 1841 they accounted only for one eighth of the then more than 3.000 students.[9] Nevertheless, the faculty of theology still occupied a distinguished place within the university. The Catholic Church was one of the pillars of the Habsburg throne and the archbishop of Prague was at the same time the chancellor of the university.

7 The basic synthetic work on the history of Charles University was published on the occasion of the last semi-round anniversary in 1998: František Kavka and Josef Petráň (eds.), *Dějiny Univerzity Karlovy* [*History of Charles University*], 4 vols. (Prague: Karolinum 1995–1998). This work also appeared in an abridged English translation: František Kavka and Josef Petráň (eds.), *History of Charles University*, 2 vols. (Prague: Karolinum 2001).

8 Jan Havránek, "The university professors and students in nineteenth-century Bohemia", in: Mikuláš Teich (ed.), *Bohemia in history* (Cambridge: University Press 1998): 219–221.

9 Jan Havránek (ed.), *Dějiny Univerzity Karlovy* [*History of Charles University*] (Prague: Karolinum 1997), vol. 3: 19.

Church supervision, together with the control of the absolutist government, resulted in a limited space for Prague academia to strive for a union of science and learning and, in general, professors did not have much opportunity for personal initiatives. Deviations from prescribed textbooks were banned and any sort of reflection on new intellectual movements within the lecture courses was unthinkable. One of the early and most extreme examples in this respect was the fate of Bernard Bolzano, an influential philosopher and mathematician, dean of the faculty of philosophy, who in 1819 was removed from his professorship because of his reformist views and his strong influence on the students.[10] Authors of the conception of the jubilee celebrations obviously had to take this political and social situation into account.

As a kind of overture to the actual preparations of the university jubilee of 1848, some publications were set up in the 1830s that focussed on the history of the university. The academic senate turned its attention in the first place to the university archives and the materials kept there.[11] Tutors at the faculty of arts began with the preparation of a large publication series on the history of Prague's *studium generale*, particularly in the form of editions of sources related to the earliest history of the university. An editorial board was set up for what eventually would become the *Monumenta historica universitatis Carolo-Ferdinandae Pragensis*. In the first half of the 1830s, the two oldest registers of the university were published, one of the faculty of arts (*Liber decanorum*) and one of the faculty of law. More than ten years later, an edition of the earliest normative texts followed, such as the oldest university statutes.[12] The series was supposed to continue with a volume containing archival documents from the fifteenth century. Sadly however, it was abandoned, mainly because after the jubilee, the interest in the history of the university largely waned.[13] Moreover, already when the anniversary drew

10 Marie Bayerová, *Bernard Bolzano. Evropský rozměr jeho filosofického myšlení* [*Bernard Bolzano. The European dimensions of his philosophical thinking*] (Prague: Filosofia – Philosophia 1994).

11 Karel Kučera and Miroslav Truc, *Archiv University Karlovy. Průvodce po archivních fondech* [*Archives of Charles University. Guide to the archival collections*] (Prague: SPN 1961): 37–38.

12 *Liber decanorum Facultatis Philospohicae Universitatis Pragensis ab anno Christi 1367 usque ad annum 1585, Pars I–II* (MHUC-FP I/1-2) (Prague: Johann Nep. Gerzabek 1830–1832), *Album seu matricula facultatis juridicae universitatis Pragensis ab anno Christi 1372 usque ad annum 1418* (MHUC-FP II) (Prague: Johann Spurny 1834), Antonius Dittrich and Antonius Spirk (eds.), *Statuta Universitatis nunc primum publici juris facta* (MHUC-FP III) (Prague: Johann Spurny [1848]).

13 According to Wáclaw Wladiwoj Tomek, *Paměti z mého života I* [*Memoirs of My Life*] (Prague: Matice Česká 1904): 183.

closer, most of the preparatory work focussed in another direction, one that was more visible to a broad public.

A committee for the preparation of the celebration was established in April 1842.[14] Each of the individual faculties was represented in the committee by two members (and an equal number of substitutes), completed with a separate representative of the senate. During the entire time of its existence, the president of the committee was Jakob Beer, professor at the faculty of theology, Grand Master of the Military Order of the Crusaders of the Red Star (*Ordo militaris crucigerorum cum rubea stella*)[15] and a former student of Bernard Bolzano. The committee immediately became very active and on its ninth meeting in November 1842, it decided that several projects should be undertaken in connection with the upcoming festivities, including a renovation of the *Karolinum*, a memorial to Charles IV, publications on the history of the university, and the design of a jubilee medal.[16] Financial means for these majestic projects were to be obtained mainly by a public fund-raising campaign. The committee not only counted on rich and important individuals, some of whom indeed decided to support the celebrations generously, but also on university graduates in the ranks of clerics, physicians, and lawyers, working outside the capital city of the kingdom. Representatives assigned by the university and, in some cases, by regional administrative and financial authorities managed negotiations with them and the actual transfer of their contributions.[17]

14 Files of the committee for celebrating the 500th anniversary of Prague University are kept in the Institute of the History of Charles University and the Archives of Charles University, collection Akademický senát 1796–1882, boxes nos. 127 and 128, inventory nos. 141–144. Between 1842 and 1849, the committee kept its own filing register (Institute of the History of Charles University and Archives of Charles University, collection Akademický senát 1796–1882, box no. 127, inventory no. 141; hereinafter indicated as Filing register of the committee for celebrating the 500th anniversary of Prague University). Further on, we refer to reference numbers from this diary.

15 In the nineteenth and early twentieth century, members of this purely Bohemian (in a territorial and not in a national sense) knight order, which was founded in 1237, were often linked to the faculty of theology and many of them served as university rectors.

16 Filing register of the committee for celebrating the 500th anniversary of Prague University, reference no. 45.

17 Hundreds of people became involved in this campaign, with contributions ranging from several dozen kreuzer up to hundreds of guilders. Institute of the History of Charles University and Archives of Charles University, collection Akademický senát 1796–1882, box no. 128, inventory no. 142: *Printed list of subscribers*.

The committee published its concept of the jubilee celebrations in a lengthy manifesto in March 1843.[18] It announced the abovementioned projects and added a number of others, which were in the end not implemented at all (e.g. setting up a scientific journal, the creation of a system of stipends for young university lecturers) or only realized later (e.g. the publication of an important treatise concerning the foundation period[19]). The announcement of the public collection campaign in the manifesto was enforced by making reference to the financial support that the archbishop of Prague and professors of the faculty of law had already promised. In line with the tradition of the university and the somewhat mythical personality of its founder, and in order to attract contributions from the whole empire (and if possible even from abroad), the committee clearly emphasized the regional and international importance of Prague's *studium generale*:

> Die Prager Universität war die erste in deutschen und slavischen Landen [sic]; ihr Wirken war bestimmt, weit über Böhmens Gränze zu reichen. [...] So verschmolz zugleich das Leben der Universität mit dem Leben der ganzen Nation; sie riefen wechselseitig ihre Geschicke hervor, und theilten sie mit einander. Die Geschichte unserer Hochschule ist unauflösbar verbunden mit der Geschichte unserer Hauptstadt und unseres Vaterlandes.

Altogether, the committee, which existed until the end of 1849, met more than forty times. When it became apparent that the renovation of the *Karolinum* was unfeasible under the given financial circumstances, the committee gradually focussed its efforts on two objectives, which both required a great deal of

18 The appeal was printed in German with 6.000 copies, and had to go through the censorship (for more on the practice of censorship, see below in connection with the preparation of Tomeks *Dějiny univerzity*). Filing register of the committee for celebrating the 500th anniversary of Prague University, reference nos. 53, 63–65, 67. One copy is kept in the Institute of the History of Charles University and Archives of Charles University, collection Akademický senát 1796–1882, box no. 127, inventory no. 142.

19 For this purpose, an old-Czech manuscript by Tomáš Štítný ze Štítného, philosopher and one of the first students of Prague University, was chosen and edited by Karel Jaromír Erben, archivist of the city of Prague. Karel Jaromír Erben (ed.), *Tomáše ze Štítného Knížky šestery o obecných věcech křesťanských* [*Tomáš Štítný's Six Books on General Christian Matters*] (Prague: K. Gerzabek 1852). The front page of the book clearly states that the manuscript was published "by the University of Prague to remember its foundation 500 years ago".

attention and time: unveiling a memorial to Charles IV and writing the history of the university.

Already at the end of 1842, the committee appointed the historian Wácslav Wladivoj Tomek,[20] lecturer at the faculty of arts and later rector of the university, to write a history of his own institution. This decision was made after it became clear that none of the four official historiographers of the individual faculties wanted to undertake this task.[21] Moreover, Tomek was recommended by František Palacký, the so-called founder of modern Czech historiography and a skilled organizer of the scientific community.[22] In order to facilitate his work, the committee arranged for Tomek to be given access to various Prague archives and the faculty historiographers provided him with the material that they had gathered up until then. And so it can be explained how this young historian – who at the same time worked on a history of the city of Prague and was involved in the preparation of an edition of the oldest Bohemian diplomatic material – was asked to compile the history of the university.[23] After several years of research and note-taking, Tomek began to write in 1845 and the first volume was finished just a year later.[24] However, the original plan to publish the work in four

---

20 Miloš Řezník (ed.), *W.W. Tomek. Historie a politika (1818–1905)* [*W.W. Tomek. History and Politics (1818–1905)*] (Pardubice: Univerzita Pardubice 2006).

21 Each faculty had its own official historiographer who was supposed to deal with the faculty's history. The historiographer was chosen from amongst the professors whose area of expertise was to some degree connected with history. During the period of preparation of the quinquecentennial, they were Johann Baptist Smutek, expert in ecclesiastical history from the faculty of theology; Georg Norbert Schnabel, statistician from the faculty of law; Karl Johann Vietz, scholar in general history from the faculty of philosophy; and Anton Jungmann, professor of obstetrics of the faculty of medicine and author of its brief history. When Tomek handed in the first part of his institutional history in the summer of 1846, these faculty historiographers were asked by the committee for the preparation of the jubilee to review it. See Filing register of the committee for celebrating the 500th anniversary of Prague University, reference nos. 170–181.

22 In the 1840s, Palacký was secretary of the Czech Museum, general secretary of the Royal Bohemian Society of Sciences, and the official historiographer of the Czech estates. And although he was not a graduate of Prague University, he had close contacts with a number of its professors. See Jan Havránek, "František Palacký a Univerzita Karlova [František Palacký and Charles University]", *Acta Universitatis Carolinae – Historia Universitatis Carolinae Pragensis* 21 (1981), no. 1: 67–81.

23 Tomek, *Paměti* (1904): 181–185.

24 In April 1846, Tomek handed in the first part of his manuscript "in beiden Landessprachen," which is both in Czech and in German. See Filing register of the committee for celebrating the 500th anniversary of Prague University, reference no. 192.

volumes in both Czech and German failed, because it soon became evident that – not too surprisingly – Tomek could not cover the entire history of the university in this format by the deadline he was given.[25] In January 1847, he was therefore asked by the committee to compile a shorter, more general summary of the university's history.[26]

In 1849, so a short time after the jubilee, the first part of Tomek's detailed history of the university was finally published in Czech. It covered a little over the first fifty years of the university's existence.[27] The shorter work was also published in 1849, albeit only in German.[28] In the introduction, the author explained that the Czech version was intended as a "pragmatic" history with a series of notes, and more volumes would follow. Yet, although Tomek did publish several detailed preparatory studies on the university's history in the sixteenth century, the planned volumes, which were to cover the entire existence of the university, were never published.[29] And whereas the Czech monograph was published without a foreword, in the German compendium Tomek explained the origins of both books.

In his introduction, Tomek also apologized for several parts of the text by saying that the work had been sent to the printers before censorship was abolished in the spring of 1848. Until then, and from 1817, Austrian censorship had been centralised in the hands of count Josef Sedlnitzky, head of the police and censorship office.[30] In general, the censors looked for anything that may threaten religion, state, or decency and morals. Even historical treatises were subject to it, as Tomek's patron Palacký was to find out when he wrote the first volumes of his innovative work *Geschichte von Böhmen* in the mid-1840s. The censors objected, for instance, to the description of the figure of Jan Hus (religious reformer and rector of Prague University at the beginning of the fifteenth century), and to the treatment of the incident with German scholars

25 Tomek, *Paměti* (1904): 204–205, 227.

26 Tomek, *Paměti* (1904): 246.

27 Wáclaw Wladiwoj Tomek, *Děje university pražské, díl I.* [*History of Prague University*] (Prague: České museum – Řivnáčovo nakladatelství 1849).

28 Wáclaw Wladiwoj Tomek, *Geschichte der Prager Universität. Zur Feier der fünfhundertjährigen Gründung derselben* (Prague: Gottlieb Haase und Söhne 1849): iii–iv.

29 Bohumil Jiroušek, "Historik W.W. Tomek [Historian W.W. Tomek]", in: Řezník (ed.), *W.W. Tomek, historie a politika* (2006): 20–21.

30 On Sedlnitzky's relationship to science and scientists, see Michal Chvojka, *Josef Graf Sedlnitzky als Präsident der Polizei- und Zensurhofstelle in Wien (1817–1848). Ein Beitrag zur Geschichte der Staatspolizei in der Habsburgermonarchie.* (Schriftenreihe der Internationalen Forschungsstelle "Demokratische Bewegungen in Mitteleuropa 1770–1850" 42) (Frankfurt am Main: Peter Lang 2010): 209–251.

who had left the university after the issue of the Decree of Kutná Hora in 1409.[31] In his dealings with the authorities, Palacký was represented by Tomek, so that by the time he wrote his university history, Tomek had already had plenty of experience of the forces of censorship, resulting in the rather cautious tone of the book.[32] Tomek's memoirs reveal that the approach towards the proposed university history by the censor – Jan Pravoslav Koubek, who was at that time also professor at the faculty of arts – was slow, but generally accommodating.[33]

The smaller, single-volume history of the university was meant to appeal to a broad range of readers and to be more accessible. This was in all likelihood also the main reason why it was published in German. The decision that the committee took in 1848 regarding the anniversary publication looked reasonable and surprisingly modern: they opted for a combination of a detailed treatment in the local language, with a more popular, accessible publication in a world language. A similar approach to anniversary publications is common to this day[34] and the same solution was also adopted for marking Prague University's jubilee 150 years later.[35]

The creation of a memorial to Charles IV was also intended to have an impact on the broader public. Gradually, it even became the main symbol of the celebration and it is still its most visible reminder. The idea of erecting a memorial to the founder of the university on the occasion of the jubilee was first formulated by Andreas Neureutter, professor at the faculty of law. Neureutter had already submitted the first proposal for celebrating the

---

31 With this decree, King Wenceslas IV gave three votes to the local Bohemian nation in the affairs of the university, against one collective vote of the remaining nations (Poland, Bavaria and Saxony). See František Šmahel, "The Kuttenberg Decree and the Withdrawal of the German Students from Prague in 1409: a Discussion", *History of Universities* 4 (1984): 153–166. It was one of the major decisions leading to the foundation of the University of Leipzig, see p. 165.

32 Jiří Kořalka, *František Palacký (1798–1896). Životopis* [*František Palacký (1798–1896). A Biography*] (Prague: Argo 1998): 222–227.

33 Tomek, *Paměti* (1904): 263–264.

34 This applies mainly to smaller linguistic communities. There are several concrete examples, such as publications connected with anniversaries of universities in Vilnius and Copenhagen: Jonas Kubilius (ed.), *Vilniaus universiteto istorija* [*History of Vilnius University*], 3 vols. (Vilnius: Mosklas 1976–1979); Jonas Kubilius (ed.), *A Short History of Vilnius University* (Vilnius: Mosklas 1979) and Svend Ellehøj et al., *Københavns Universitet 1479–1979* [*University of Copenhagen, 1479–1979*], 14 vols. (Copenhagen: Gads Forl. 1979–2005); Svend Erik Stybe, *Copenhagen University. 500 Years of Science and Scholarship* (Copenhagen: DK Books 1979).

35 See footnote 7.

university's quinquecentennial to the academic senate in November 1840, in his position as dean of the law faculty. According to his vision, the celebrations should be preceded by a thorough reconstruction of the *Karolinum*. As noted above, the committee for the preparation of the jubilee had initially agreed with this idea. Sadly however, it soon became apparent that funds for a reconstruction and alteration of the main university building were lacking.[36] Therefore, the organizers focussed on building a memorial to Charles IV.[37]

The selection of a suitable location and artist were subject to lengthy negotiations. The fact that the committee was presided over by Beer, Grand Master of the Military Order of the Crusaders of the Red Star, as mentioned before, greatly facilitated the talks concerning the location. The ideal location that was suggested by the committee, situated at the end of the Charles Bridge with a view upon Prague Castle, also happened to belong to this Knight's Order. An agreement was reached in 1845, allowing the site chosen for the memorial to be leased to the university permanently, although the committee still had to negotiate additionally with the Prague magistrate regarding permission for requisite building alterations. More controversy within the committee surrounded the choice of the statue's artist. In consultation with the Prague Society of Patriotic Friends of Art, the original idea to entrust the order to a local artist was quickly abandoned. The view prevailed that a sufficiently qualified one had to be sought abroad, more so after the decision was taken on the material of the memorial, being bronze or some iron alloy. Finally, in the course of 1843, through the agency of Christian Ruben, director of the Academy of Arts in Prague, the committee accepted the offer of Ernst Julius Hähnel, sculptor and professor at the Academy of Arts in Dresden.

Looking back, commissioning the design of the memorial to a young but already respected artist was one of the crucial decisions connected with the jubilee. Hähnel supported the choice of the memorial's location because of its closeness to and link with other buildings from the time of Charles IV. Regarding the general conception of the memorial, he also accepted the basic

36 Petráň, *Karolinum* (2010): 50–51 and Miroslav Kunštát, "Stavební a umělecké proměny Karolina v 19. a 20. století [Changes to the Building and Art of *Karolinum* in the 19th and 20th Century]", *Acta Universitatis Carolinae – Historia Universitatis Carolinae Pragensis* 26 (1986), no. 2: 88.

37 The following two paragraphs are based mainly on Miroslav Kunštát, "Monumentum fundatoris. Pomník císaře Karla IV. k 500. výročí založení pražské univerzity [Monumentum Fundatoris. Memorial to Emperor Charles IV at the Occassion of the 500th Anniversary of Foundation of the Prague University]", *Acta Universitatis Carolinae – Historia Universitatis Carolinae Pragensis* 40 (2000), no. 1–2: 39–51.

idea proposed by the committee: a dominant, larger than life-size statue of the emperor with the founding charter of the university in his hand, together with four smaller, allegorical statues symbolising the individual faculties and four statues of Charles' contemporaries. In January 1844, the first model of the memorial was completed. Hähnel personally brought it to Prague and successfully presented it to a committee of experts and later also to a broader public. In the course of the following year, the committee finally managed to secure crucial permissions from all relevant authorities and consent of the ruling emperor, Ferdinand I. The actual building of the memorial itself then continued more or less according to schedule. A depiction of it was later also used for the obverse of a commemorative medal designed by the engraver Wenzel Seidan (see figure 2.2).[38]

FIGURE 2.2 *Commemorative medal issued on the occasion of the university's quinquecentennial in 1848.*
PICTURE COURTESY OF THE INSTITUTE OF THE HISTORY OF CHARLES UNIVERSITY AND THE ARCHIVES OF CHARLES UNIVERSITY

38 Otto Herber, *Insignie, medaile, taláry Univerzity Karlovy* [*Insignia, Medals, and Gowns of Charles University*] (Prague: Univerzita Karlova 1987): 75–77.

Apart from having to deal with the financial issues, the memorial as a whole was probably the most demanding task the preparatory committee had to deal with from the beginning. The date by which it was supposed to be finished even played a crucial role in the timing of the celebrations. At the end of 1847, it was already clear that the memorial would not be finished by 7 April 1848, being the anniversary of the founding charter. It was therefore decided that the jubilee should be postponed to late September – early October 1848.

According to the programme, representatives of the university and all the guests would assemble on 29 September shortly before noon in the great hall of the *Karolinum*, listen to the rector's speech in Latin, and afterwards walk together in procession to the Church of Our Lady Before Týn.[39] Representatives from all German, Austrian, Swiss, and most other important European universities would be invited to join the festivities.[40] Led by the archbishop of Prague, the procession would move from the church to the new memorial of Charles IV, where Beer, president of the jubilee committee, would hold a speech in German. The memorial would be unveiled, consecrated, and a celebratory song would be struck up in both national languages. On the following day, a graduation ceremony of honorary doctors would be organized in the presence of the archbishop, who would personally present the diplomas. A festive meeting of the Royal Bohemian Society of Sciences would follow and the day would come to an end with a great ball. After a break of two days, the celebrations would continue with the inauguration of the new rector, again in the Church of Our Lady Before Týn. In the evening, a torchlight procession of the students to the memorial of Charles IV and the *Karolinum* would conclude the celebrations.[41]

The Anniversary in the Shadow of the 1848 Revolution

However, between April and September 1848, social and national movements of the Spring of Nations also reverberated in the Czech Lands, interfering with

39 This is the most important church in Prague's Old Town close to the *Karolinum*, hosting most of the masses on the occasion of all kinds of university events.

40 In particular, Paris, Bologna, Naples, Cambridge, Oxford, Uppsala, Rome, Dublin, Leuven, Brussels, Florence, Copenhagen, Leyden, Edinburgh, Dorpat, Moscow, Oslo, Saint Petersburg, and London. Institute of the History of Charles University and Archives of Charles University, collection Akademický senát 1796–1882, box no. 128, inventory no. 142: *Undated list of foreign invitees.*

41 Institute of the History of Charles University and Archives of Charles University, collection Akademický senát 1796–1882, box no. 128, inventory no. 142: *Proposal of the jubilee programme, signed by vice rector Josef Reisich* (29/12/1847).

the jubilee preparations. The upheaval had begun in Paris in February, with demonstrations causing the abdication of Louis Philippe I. Following the Paris' example, uprisings occurred in Vienna in early March, leading to the overthrow of State Chancellor Metternich's government and the subsequent adoption of a more liberal policy. In the street skirmishes that preceded Metternich's resignation, Viennese students played a key role.

News of what had occurred in the imperial capital soon arrived in Prague, where it was received with much interest. Prague students immediately sympathized with their Viennese comrades, forming a petition movement and even joining forces in armed Academic Legions.[42] The *Karolinum* gradually became one of the symbols of the revolution. Already on 15 March 1848, it was the stage of a student assembly that adopted a petition demanding, among other measures, the unlimited freedom of teaching, the possibility of visiting foreign universities during the course of study, and the equality of both languages of the region, Czech and German. The presence of the mayor of Prague – being the most senior representative of state power in the town – and his speech indicated that it was not possible to ignore the students as a social group and that they had to be taken seriously. Students clearly represented the most radical part of the revolutionary movement. More than half of the names on the list of "politically compromised individuals", which was compiled by Prague police headquarters in the mid-1850s, consisted of students (38 out of 72).[43]

In this atmosphere of revolutionary excitement, students were in the end the only ones to celebrate the anniversary. Their celebration took place in the *Karolinum* courtyard on 7 April 1848, exactly on the day of the 500th anniversary of the signing of the foundation charter. The attendance was high and in addition to traditional elements, such as a speech in honour of Charles IV, the festivity also included revolutionary elements that were added on the spot. To the sound of funeral music, a top hat and a high collar were burned on an improvised bonfire, and the nearby Jezuitská [Jesuit] Street was renamed Karlova [Charles] Street (the street has retained its name change up to this day). Both actions had a highly symbolic significance. For the participants they were an expression of their opposition to the absolutist, extremely clerical regime. The popularizing

42 The role of the university in the revolutionary events of 1848 is described in detail by Jan Havránek, "*Karolinum* v revoluci 1848 [*Karolinum* in the Revolution of 1848]", *Acta Universitatis Carolinae – Historia Universitatis Carolinae Pragensis* 26 (1986), no. 2: 35–75.

43 Petra Ševčenková, "Edice soupisů tzv. politicky kompromitovaných pro Prahu z roku 1857. Neoabsolutistická perzekuce v Čechách ve světle úřední korespondence [The edition of lists of so-called politically compromised individuals in Prague 1857. Neo-absolutist persecution in Bohemia in the light of official correspondence]", *Praginae historiae* 14 (2006): 461.

atmosphere of the ceremony was emphasized by the performance of a brass band and the donation of ten barrels of beer, rolled out on the orders of the rector in support of the impecunious students.[44]

In his memoirs, Tomek, the university historiographer, noted with some derision that under different circumstances the university celebrations

> [...] would have interested all classes of the population and would have been very grandiose. However, in the excitement of the time, it turned into a mere student celebration, which passed by in a rather undignified manner in a courtyard of the *Karolinum*. Indeed, the rector delivered a speech [...], but he did so in front of a very restless audience.[45]

In the late spring of 1848, a further radicalization of sentiments led to armed clashes. In mid-June, the army occupied all strategic points and General Alfred Windischgrätz, head of the Bohemian regiment, ordered the artillery even to shell the city. The violent clashes in the streets of Prague also affected the buildings of the university. The army took control of the *Karolinum* at the beginning of the conflict, after it had been occupied by some sixty students, supported by several dozen craftsmen and labourers. In the course of the brief struggle, the troops had damaged the university archives, from which the golden bull of the foundation charter of Charles IV disappeared. (Fortunately, it was later recovered.)[46] In the end, the revolution suffered a decisive defeat.

During this time, attempts to organize the postponed anniversary celebration naturally came to a halt. When the situation had calmed down a bit, the preparatory committee first of all, "in Anbetracht der gegenwärtigen Zeitenverhältnisse und itzt ganz geänderten Umstände", decided to shorten the official celebrations to one day. The programme was to be adjusted accordingly.[47] The situation in Prague was, however, once again affected by the

---

44 Josef J. Toužimský, *Na úsvitě nové doby. Dějiny roku 1848 v zemích českých* [*At the Dawn of a New Age. History of 1848 in the Czech Lands*] (Prague: Josef R. Vilímek 1898): 309–310.

45 Tomek, *Paměti* (1904): 266.

46 It remains unknown who stole Charles's *bulla aurea* during the fight for the *Karolinum*. The bull returned to the university's possession in January 1854, from an antique collection of the deceased Prague collector Josef Pachl. Bohdan Zilynskyj, "Příběh ukradené a znovunalezené buly Karla IV. (1848–1854) [Story of the Stolen and Recovered Bull of Charles IV (1848–1854)]", *Acta Universitatis Carolinae – Historia Universitatis Carolinae Pragensis* 41 (2001), no. 1–2: 49–58.

47 Institute of the History of Charles University and Archives of Charles University, collection Akademický senát 1796–1882, box no. 128, inventory no. 142: *Notes from a meeting of the Committee for Celebrating the 500th Anniversary of Prague University* (14/07/1848).

developments in the imperial capital, where unrest turned into an all-out armed uprising. On 12 October 1848, a committee of Prague burghers addressed an appeal to the preparatory committee of the university jubilee with the request to postpone the festivities in view of the "Oktober-Ereignisse" in Vienna. The preparatory committee passed on this request to the academic senate, which on 18 October decided to meet the demand. The celebrations were postponed without setting a new date and this decision was to be announced in the daily press.[48] In the end, the ceremony did not take place at all. The memorial was quietly unveiled on 31 January 1849[49] (see figure 2.3) and honorary doctorates were conferred in a similar manner, i.e. the diplomas were sent by mail or delivered by hand to the laureates.

So, the preparations to celebrate the 1848 jubilee took six years to complete. The result, however, bore little resemblance to the original plans, at least as far as the proper celebrations were concerned. When evaluating the jubilee by making use of Thomas P. Becker's typology, which distinguishes four constitutive parts of nineteenth-century university jubilee celebrations,[50] it can be concluded that the original conception was almost completely turned inside out:

(1) Popular celebration. The public unveiling of the memorial to Charles IV was intended as such, but in reality, the student meeting on 7 April 1848 in the *Karolinum* courtyard was the only part of the jubilee that included the larger public, and then even in an improvised manner.

(2) Student celebration. This meeting was thus de facto a combination of a student party and a popular celebration. The actual role of the students contrasts sharply with the part that was originally reserved for them, according to the documents of the preparatory committee. In the long-term plans of the organizers, student participation featured only quite marginally.

48 Filing register of the committee for celebrating the 500th anniversary of Prague University, reference nos. 246–248.

49 Kunštát, "Monumentum fundatoris" (2000): 50; Alois Kubiček, Alena Petráňová and Josef Petráň, *Karolinum a historické koleje University Karlovy v Praze* [Karolinum *and historical colleges of the Charles University in Prague*] (Prague: SNKLU 1961): 96–97. The latter book, which was written under difficult political circumstances in the late 1950s and early 1960s, voices high regard for the statue: "We are grateful for the fact that the anniversary year left us the distinguished statue of Charles IV, being the work of a representative of the German Romantic school and a statue of outstanding artistic merit, which became part of Prague both due to its standard and its location."

50 Becker, "Jubiläen als Orte universitärer Selbstdarstellung" (2008): 92–93.

FIGURE 2.3 *Memorial to Charles IV.*
PICTURE COURTESY OF THE INSTITUTE OF THE HISTORY OF CHARLES UNIVERSITY AND THE ARCHIVES OF CHARLES UNIVERSITY

(3) Academic celebration. The typical solemn ceremony, in which congratulatory addresses were delivered and honorary doctoral degrees were awarded, in the end did not take place, although it was clearly included in the original plans.

(4) The jubilee as a cultural event. The same applies to musical, literary and dramatic performances. They were scheduled by the preparatory committee, yet never took place.

These utterly disastrous celebrations contrast with the successful (although, in almost all the cases, postponed) completion of long-term projects, which were supposed to be presented to the public during the jubilee year: the jubilee medal, various publications, and, above all, the memorial to Charles IV.

From Charles-Ferdinand University to the Czech Charles University (1848–1945)

Something similar happened a hundred years later. The anniversary celebrations of 1948 were also largely altered due to the political circumstances of the time. During the centenary between the two jubilees, the position, outlook and size of the celebrating university in Prague had changed profoundly. One of the consequences of the revolutionary events of 1848 was a gradual increase in the national and linguistic tensions between the Czech- and German-speaking populations. This also had a large impact on the university, which was eventually split into a Czech and a German section in 1882, each with four faculties that coexisted in parallel. The new institutions were connected by a shared past (they were both legal successors of Charles's original *studium generale*), by the alternate use of the original insignia and the oldest university buildings, and by the common university archives.

However, these common elements became the subject of a series of disputes, which culminated in the mid-1930s. After the creation of an independent Czechoslovak state in 1918, Czech national circles more or less successfully tried to suppress the rights of the German university as being one of the two heirs of the original medieval university. In 1920, the Czech university designated itself *Universita Karlova* [Charles University], dropping the Habsburg name Ferdinand, whereas its German counterpart was simply given the name of *Die Deutsche Universität zu Prag*. The controversy continued over the following years and, in 1934, resulted in the act "On the relation between Prague universities". It stipulated that the old insignia and other historical artefacts that symbolized the academic tradition (including the university archives) came into the exclusive possession of the Czech university. Of course, representatives of the German-speaking university resisted and tried to prevent these measures, largely in vain. The continuous dispute was keenly followed by a broad public, since it clearly had a larger meaning and was not limited to the university alone.[51]

51 For a clear overview of the relationship between the two universities, see Hans Lemberg, "Die tschechische Universität in Konkurrenz zur deutschen Universität (1882–1939)", in:

The conflict between both universities mainly remained one of an ideological character. From the end of the nineteenth century, faculties of both the Czech and the German university received a number of new buildings with modern equipment.[52] These improved facilities of the individual faculties sadly contrasted with the state of the *Karolinum*, which was jointly used by both universities. In fact, the *Karolinum* was divided into a Czech and a German part, with separate routes of access so that members of the two academic communities would meet each other as little as possible. Only the great hall with an irreplaceable symbolic significance was used by both institutions in turn. In result of the law of 1920, the *Karolinum* became the exclusive possession of the Czech university, but its German counterpart still had the right to use it, until a new building was completed or until another adequate substitute location was found, something which did not happen during the interwar years.

Various attempts to reconstruct the *Karolinum* were undertaken from the mid-1930s. Already in 1934, the Czech university established a committee for the renovation of the *Karolinum* on the initiative of Rector Karel Domin, who was known for his somewhat nationalistic persuasion.[53] Domin also called for a "national fund-raising for the reconstruction of the *Karolinum*", a proposal which was enthusiastically supported by the students.[54] To launch the campaign, he published the representative publication *Karolinum – A national possession*, together with two colleagues.[55] To some extent, all these initiatives added fuel to the conflict between Czechs and Germans. As a crucial place of remembrance in Prague, the oldest university building played a prominent role in the national disputes between both groups in the latter half of the 1930s. During Domin's rectorate, a systematic evaluation of the construction

---

Blanka Mouralová (ed.), *Die Prager Universität Karls IV. Von europäischen Gründung bis zur nationalen Spaltung* (Potsdam: Deutsches Kulturforum östliches Europa 2010): 157–185.

52 For more details on academic architecture in Prague in that period, see Michaela Marek, *Universität als Politikum. Die Repräsentationsbauten der Prager Universitäten 1900–1935 und der politische Konflikt zwischen 'konservativer' und 'moderner' Architektur* (München: Oldenbourg 2001).

53 Some members of the committee (e.g. Josef Cibulka and Václav Vojtíšek) played a key role in the preparations for the post-war anniversary celebrations.

54 A detailed, though rather one-sided account of the activities regarding the *Karolinum* can be found in Karel Domin, *Můj rektorský rok. Z bojů o Karolinum a za práva Karlovy university [My year as rector. On the struggles for the* Karolinum *and the rights of the Charles University]* (Prague: A. Neubert 1934).

55 Karel Domin, Václav Vojtíšek and Josef Hutter (eds.), *Karolinum, statek národní [Karolinum – A national possession]* (Prague: L. Mazáč 1934).

and history of the building began. Preparations for a reconstruction of the *Karolinum* continued until the beginning of the war. Even during the war, the work went on, yet it passed under the management of the German university.

The upcoming anniversary highlighted not only the need for a reconstruction of the *Karolinum*, but also the necessity of a modern synthetic treatment of the university's history. In 1935, a group of senior lecturers and young professors of the Czech faculty of arts, led by Václav Vojtíšek, submitted a proposal for a compilation of a history of the university and the publication of a university *diplomatarium* (the edition of a set of university privileges and other similar medieval documents). However, their older colleagues contested the proposal, in particular the renowned professors Josef Pekař and Josef Šusta. Their negative attitude was a result of both professional and generational rivalry. They recommended partial alterations, such as the collaboration with particular specialists in the history of natural sciences, and passed the amended proposal for deliberation to the appropriate committee of the academic senate. The start of the actual work was delayed mainly by a combination of plodding bureaucracy slowness and a deteriorating foreign and domestic political situation.[56]

After the occupation of the Czech Lands by Hitler in his attempt to create a Great Germany and after the establishment of the Protectorate of Bohemia and Moravia, the German-speaking university was renamed *Die Deutsche Karls-Universität* in September 1939.[57] Two months later, on 17 November, all the Czech-speaking universities and institutes of higher education were closed down, nine leaders of the student movement were executed, and hundreds of other students were sent to the Sachsenhausen concentration camp. Most of the property of the Czech university was taken over by its German counterpart, which at the same time became officially one of the universities within the German Reich. Its representatives planned the celebrations of a centenary jubilee fully in the spirit of the ruling Nazi regime. The pompous architecture of the Third Reich also influenced the plans for a reconstruction of the *Karolinum*, which was to be directed by the architect Josef Zasche.[58] Given the

56 Institute of the History of Charles University and Archives of Charles University, collection Charles University Faculty of Arts, box no. 5, inventory no. 52: *Reports and minutes* (05/12/1935 and 30/01/1936).

57 On the German Charles University, see Alena Míšková, *Die Deutsche* (*Karls-*) *Universität vom Münchner Abkommen bis zum Ende des Zweiten Weltkrieges* (Prague: Karolinum 2007).

58 Kunštát, "Stavební a umělecké proměny Karolina" (1986): 96–97.

developments on the battle fronts of the Second World War, however, the preparations of the jubilee as well as the rebuilding of the *Karolinum* had to give way to other, more pressing issues.

In 1945, the situation was reversed. In post-war Czechoslovakia, there was no longer a place for a German university alongside the restored Czech Charles University. The German university was officially declared defunct by a decree of the President of the Republic issued on 18 October 1945, its effect even retroactively going back to 17 November 1939. The 1948 celebrations thus became a purely Czech affair. During the anniversary year, former professors of the German Charles University commemorated their Prague alma mater separately with ceremonial lectures, which were part of the celebrations at the newly established Adalbert-Stifter-Verein.[59]

Preparations for the 1948 Anniversary of the Charles University as a Czech National Celebration

The short period known as the "Third Czechoslovak Republic" – which lasted from the liberation in May 1945 until the communist takeover in February 1948 – can be characterized as a time of limited democracy and a sharp general shift towards the left. When academic freedom was questioned by the communist Minister of Information Václav Kopecký in the autumn of 1947, university representatives came out openly to defend it. Like the rest of society, Czech students too were politically polarized, although those of a democratic inclination predominated. Since the summer of 1945, institutes of higher education were overwhelmed by applicants, who in some cases had to wait for more than five years (at some faculties of the Charles University,

59 *Die deutsche Universität in Prag. Ein Gedenken anläßlich der 600 Jahrfeier der Karls-Universität in Prag* (Gräfelfing bei München: Edmund Gans 1948). This brochure includes the lecture "Alma mater Pragensis" (given by the legal historian Wilhelm Weizsäcker on 30 September 1948 in Munich) and the lecture "Die Prager Universität und das Schicksal Mitteleuropas" (given by the historian Eugen Lemberg on 3 October 1948 in Augsburg). The original plan to commemorate the anniversary of Prague University on 18 April 1948 in the big auditorium of Munich University was abandoned after an intervention by the American civilian administration. See Miroslav Kunštát, "Centra a periférie českých Němců ve vědě a vzdělávání [Centres and peripheries of Czech Germans in science and education]", in: Kristýna Kaiserová and Miroslav Kunštát (eds.), *Hledání centra. Vědecké a vzdělávací instituce Němců v Čechách v 19. a první polovině 20. století* (Ústí nad Labem: Albis International 2011): 48.

the number of students had almost doubled in comparison with the pre-war figures).

Against this background, the preparations got under way for the 600th anniversary that was rapidly approaching.[60] Because several other anniversaries of nationwide importance fell within this year (e.g. the 11th Sokol Slet, the centenary of the Pan-Slav Congress in Prague, and the centenary of the Kremsier Parliament[61]), it was soon decided that the efforts should be coordinated. For this purpose, in June 1946 an interdepartmental committee was established, where alongside the Charles University, various ministries, cultural institutes, and research establishments were represented. Minutes from the meetings of this committee indicate that the university celebrations were seen as one of the most important events of the year. These records also show that the interests of the university were defended in the meetings of the committee by several of its members who took part in the debates as representatives of learned societies. The university archivist and chairman of the university committee for the preparation of the celebrations (see below), Václav Vojtíšek, was

60 Unless stated otherwise, the following part is based on documents in the following archival collections (this applies in particular to the minutes from the meetings of the anniversary committees): Institute of the History of Charles University and the Archives of Charles University, collection Jubilejní oslavy Univerzity Karlovy, boxes nos. 1–2.

61 The sport association Sokol was founded in 1862, inspired by the idea of combining physical fitness education with Czech national awareness. Gradually Sokol became one of the largest associations in the Czech Lands and its regular festivals, which took place in Prague from 1882, developed into central events of nationwide importance, see Claire E. Nolte, *The Sokol in the Czech Lands to 1914: Training for the Nation* (New York: Palgrave Macmillan 2002). The Pan-Slav Congress took place in Prague between 2 and 12 June 1848. It was attended mainly by representatives of Slavic nations living in the Habsburg monarchy and was meant to be a show of resistance to German nationalism. The Kremsier Assembly has to be put in the same context of the revolutionary year 1848. It refers to a session of the Imperial Constitutional Assembly of the Habsburg Monarchy, which first convened in Vienna, but later resided in the Moravian town Kremsier. Its main result was the Kremsier Constitution that featured many progressive reforms including forming a constitutional monarchy, creating a parliament that would share power with the Emperor, abolishing the privileged status and all titles of the Catholic Church within the Empire, deriving the Emperor's power from the people rather than the Grace of God and finally, making all languages and nationalities equal in the eyes of the Monarchy. However, the assembly was dismissed in March 1849 and its constitution was replaced by the reactionary March Constitution. For more on the events of 1848–1849 in the Czech Lands, see Otto Urban, *Die tschechische Gesellschaft 1848–1918. Band I* (Wien: Böhlau 1994).

elected to the national committee as a representative of the Royal Bohemian Learned Society, and another member of the university committee, the classical philologist Antonín Salač represented the Czech Academy of Sciences and Arts. The university was thus well informed about the plans for the nationwide celebrations, even though its official representative, the philosopher Josef Král, showed little interest in the work of the committee and often failed to attend its meetings.

As indicated, the academic senate of the Charles University soon established its own preparatory committee headed by Vojtíšek, which for the first time met in May 1946. It included not only professors from the faculty of arts, who had already been involved in the preparation of the jubilee before the war, but also representatives of other faculties. Later, some other academic dignitaries and specialists who were assigned specific tasks were also co-opted. Throughout the nearly two years of the committee's existence, it had in average about fifteen members. In 1947, the constantly increasing number of tasks it was facing resulted in the creation of four subcommittees, dealing with: (a) the programme, (b) the organization, (c) publications and (d) finances. A fifth subcommittee was added later on for economic and technical affairs. Because the rectorate did not have enough staff to deal with the complex agenda of the anniversary celebrations, an independent office was established in July 1947 to take care of these matters, led by Josef Hutter, musicologist at the faculty of arts of Charles University. The office of the university bursar managed financial affairs. This intricate structure behind the preparation of the jubilee clearly shows its perceived importance and scale.

The original pre-war plans to prepare a representative history of the university were eventually dropped, due to lack of time and resources.[62] Various proposals ultimately led to the decision to publish several brochures on the history of the university and its buildings,[63] a recently discovered text by

---

62 Unfortunately, the Second World War affected the university archives to a considerable extent. A large part of the oldest written documents up to the nineteenth century, as well as the original university insignia, disappeared without a trace. Václav Vojtíšek, "O archivu University Karlovy a jeho ztrátách [On the archives of Charles University and its losses]", *Archivní* časopis 1 (1951): 86–93; Karel Hruza, "Der Deutsche Insignien- und Archivalienraub aus der Prager Universität 1945", *Bohemia* 48 (2008): 349–411.

63 The English versions of these publications are: Václav Chaloupecký, *The Caroline University of Prague. Its Foundation, Character, and Development in the Fourteenth Century* (Prague: The Caroline University 1948); Václav Vojtíšek and Dobroslav Líbal, *The Carolinum: Pride of the Caroline University* (Prague: The Caroline University 1948). In addition to these, Vojtíšek published several smaller brochures in his own name.

Jan Hus,[64] and a large collective volume that would focus on the present and recent past of the Charles University. This monograph, titled *Universitas Carolina Pragensis*, in the end was never realised, even though by the beginning of 1948 the editing work was already at a rather advanced stage. Its print run was supposed to be the same as for the other official anniversary publications (with the exception of Hus's *Quodlibet*): 1.300 Czech, 700 English, 500 French, and 300 Russian copies. The number of copies in foreign languages clearly reflects both the political orientation of the university administration immediately after 1945 and the expected interest in the university jubilee by domestic and foreign universities. After the communist takeover in February 1948, it was of course unthinkable that the Russian edition would be the smallest one. The complete absence of a German edition also says a great deal. In addition to these official books, a number of other publications about the university and its history were published in connection with the anniversary, but not on the initiative of the university administration.[65]

One of the main issues the university committee had to deal with was construction work on the oldest university building. The government released considerable funds for the reconstruction of the *Karolinum* to ensure that the great hall would at least be ready on time. A somewhat smaller sum was set aside for the reconstruction of another important university building, the aforementioned *Klementinum*, which was at that time mainly used as the seat of the university library.

Several of the largest, most distinguished, and most prestigious sites in Prague were considered as alternative locations for the introductory part of the celebrations, in case the *Karolinum* was not completed on time. In particular, the National Theatre, the *Rudolfinum*, and the Vladislaus Hall in the Prague Castle were all considered good alternatives.[66] These factors

64 The committee justified the publication of the book by claiming that "for the scientific world, which is interested in the personage of the Czech reformer, a new publication would be a significant event." *Minutes from a session of the Committee for the 600th Anniversary of the Charles University* (27/05/1947). See Bohumil Ryba (ed.), *Magistri Iohannis Hus Quodlibet. Disputationis de Quodlibet Pragae in Facultate Artium Mense Ianuario Anni 1411 habitae Enchiridion* (Prague: Orbis 1948).

65 These included several other studies on Jan Hus and editions of his works, a series of popularising brochures by Vojtíšek dedicated to various subjects about the history of Prague University, several studies focussing on the foundation of the university, and the like. A summary of these works can be found in Marie Haasová-Jelínková, "Publikace k šestistému výročí založení Karlovy university [Publications on the occasion of the 600th Anniversary of Charles University]", *Český časopis historický* 48–49 (1947–1948): 469–477.

66 For the selection of the right location, the organizers had to bear in mind not only the size of the halls, but also the symbolism of the place in question. The National Theatre,

also largely influenced the planning of a date for the university celebrations. Two options were seen as possible. Firstly, there was the day on which Charles IV had published the university's privileges, i.e. 7 April. This date offered the additional advantage that the university jubilee would be the first great celebration of the year. Secondly, the pros and cons were considered regarding the postponement of the jubilee till the autumn. It would afford the organizers more time to complete the construction work and, in this scenario, the university jubilee would be the culmination of a series of anniversaries throughout 1948. In the end, taking into account the associated symbolism, as well as the argument that towards the autumn the public would already be tired of festivities, 7 April was chosen, and this decision gradually gained official support.

Salač composed a memorandum on the reconstruction of the hall at the *Karolinum*, which he presented to Prime Minister Klement Gottwald (later the first communist president) on 12 June 1946 in his capacity as representative of the national coordination committee. On this occasion, Gottwald declared that "he would do everything to make sure the work was finished by April 1948, i.e. in time for the 600th anniversary of the founding charter of the Charles University."[67] The reconstruction itself was, after the Second World War, entrusted to Jaroslav Fragner, architect and professor at the Academy of Arts. This member of the interwar cultural avant-garde and designer of a number of functionalist buildings increasingly focussed his career on the reconstruction of historical buildings. He approached the possible structural alterations of the *Karolinum* with great reverence and sensitively inserted new elements into the older parts, which have been preserved in the existing complex of buildings.[68] From the perspective of the forthcoming anniversary, the most important part of the building alterations was undertaken in the great hall, which was extended into its current form.[69]

---

opened in 1881, was the embodiment of Czech cultural emancipation of the nineteenth century. *Rudolfinum*, whilst on the same river bank, only a little newer and boasting one of the most representative concert halls in the country, lacked this clearly defined national symbolism. The late medieval Vladislaus Hall at the Prague Castle, formerly the place of coronation feasts, and since 1934 also the place where the president was elected, seemed to be a space reserved for truly extraordinary events.

67 *Minutes of the 3rd meeting of the Coordination Committee for Jubilee Celebrations of 1948* (03/09/1946).

68 For more on his involvement in the reconstruction of *Karolinum*, see Kunštát, "Stavební a umělecké proměny Karolina" (1986): 98–104.

69 Kubiček, Petráňová and Petráň, *Karolinum* (1961): 104–106.

In addition to this reconstruction work, leading Czech artists, including professors from the Academy of Arts, were approached to prepare various other works of art on the occasion of the jubilee. Most of them emphasized the traditional connection between the university and Charles IV, which was reflected in various depictions of the university's founder. The most important of these works of art were the statue of Charles IV and a tapestry with the motif of the oldest university seal and symbols of the four original faculties. Both of them still grace the great hall of the *Karolinum*. Charles IV was also depicted on the diplomas for the honorary doctors and on two jubilee medals, where he was accompanied by images of the *Karolinum*. The national aspect of the university tradition was to be embodied in the statue of Jan Hus, placed in the courtyard of *Karolinum*. The medieval rector and religious reformer is captured as he announces the Decree of Kutná Hora, which strengthened the Czech character of the university.[70]

Despite the post-war austerity measures, the National Bank of Czechoslovakia released 100 kilograms of silver to mint commemorative coins, giving another indication of the national character of and support for the jubilee. A voucher for one thousand metres of black cloth for the gowns and five metres of leather to make pouches for the diplomas had to be secured directly from the Ministry of Internal Trade. And finally, a special issue of stamps was released with the old university seal, on which Charles IV commends the university and its founding charter to St. Wenceslas, the patron saint of Bohemia. This familiar motif was also used as the general logo of the celebrations (see figure 2.4).[71]

In line with the design of the jubilee as a national celebration, the general public also needed to be involved. They were offered the opportunity to learn about the history of the university in two exhibitions. The first of them was set

---

70 See p. 30.

71 Many questions related to the origin and original purpose of this medieval seal have not yet been answered satisfactorily. Nevertheless, it was already an integral part of Prague University's tradition from the early modern period, and during the nineteenth century the motive of a ruler kneeling in front of St. Wenceslas was often paraphrased on various occasions. In 1848, a stone shield with this depiction was placed under the windows of the assembly hall on the first floor of the *Karolinum* in the direction of Železná Street. Petráň, *Karolinum* (2010): 51, 82–83. On the origin of the seal, see František Šmahel, "Das Rätsel des ältesten Prager Universitätssiegls", *Bohemia. Zeitschrift für Geschichte und Kultur der böhmischen Ländern* 43 (2002): 89–115. The fact that the seal currently appears on the reverse side of the Czech hundred Koruna note (with a portrait of Charles IV on the opposite side), illustrates the status of the seal as a lasting symbol of which the importance transcends the world of academia.

FIGURE 2.4 *Logo of the 1948 anniversary.*
PICTURE COURTESY OF THE INSTITUTE OF THE HISTORY OF CHARLES UNIVERSITY AND THE ARCHIVES OF CHARLES UNIVERSITY

up jointly in the city library and the university library at the *Klementinum*, showing manuscripts and old prints related to the university, which were usually hidden from the public. The other exhibition presented medieval documents from the university archives, old depictions of university buildings, results of analysis of the historical construction of the *Karolinum*, and models of statues of Charles IV that were created in connection with the jubilee. It was hosted by the Museum of Arts and Crafts and arranged in collaboration with Vojtíšek (see figure 2.5).

Gradually, the preparations for the anniversary celebrations drew to a close. Two issues with a certain political impact still had to be settled: the election of a new rector for the anniversary year and the selection of foreign invitees and honorary doctors. With regard to the latter, special emphasis was put on the participation of those universities which Charles University had traditionally good relationships with, and of those which had provided assistance during the Nazi occupation. The main ceremonial speech by a foreign university representative was to be given by someone from the Sorbonne, as this university had been the main source of inspiration for the foundation of the Charles University. A representative from Oxford University was to receive the honour of delivering a speech in the name of the foreign honorary doctors, because in 1943–1944 over forty Czech students had been given the chance of graduating at

FIGURE 2.5 *Opening of the exhibition on 5 April 1948. Rector Bohumil Bydžovský and the picture of the lost foundation charter.*
PICTURE COURTESY OF THE INSTITUTE OF THE HISTORY OF CHARLES UNIVERSITY AND THE ARCHIVES OF CHARLES UNIVERSITY

Oxford instead of at the Charles University.[72] Because the capacity of the *Karolinum* was limited and the number of expected guests was large, after lengthy discussions it was decided that, as a rule, universities as institutions would be invited to the celebrations – and not particular notable individuals. The selection criteria of which universities had to be invited were hotly debated. In the end, the idea prevailed that as many European universities as possible should be welcomed – with the exception of the German ones. Due to the restricted capacity of the hall, it was agreed upon to limit then the number of invitations to learned societies and academies.

Even more difficult proved to be the election of the new rector for the anniversary year. Immediately after the war, the pre-war procedure still applied,

72 This particularly concerned physicians serving in Czechoslovak foreign units. See the memoirs of one of them: Karel A. Machacek, *Escape to England* (Sussex: Book Guild Ltd. 1988).

whereby the academic senate elected a rector for one year, with different faculties filling the position in turn. In the academic year 1947–1948, it was the turn of the faculty of law. Although Karel Engliš, a famous economist, pre-war government minister, and governor of the National Bank, had the full support of the academic senate and at election received the full number of votes, his candidacy was not universally welcomed by his own faculty. Especially the Communist Party followers among the teaching staff of the law faculty opposed it. However, in view of the upcoming anniversary, his personal prestige and experience played a large role and it was hoped that his authority would speed up and simplify the bureaucracy of the anniversary committee. Even before his official acceptance of the position of rector, Engliš was energetically involved in the organization of the anniversary jubilee. He was well acquainted with all of the essential aspects of the planned celebrations and proposed several additions to the events. The most important of his contributions at this point were probably his successful negotiations with President Edvard Beneš to obtain the support of the head of the state for the restoration of Charles IV's lost foundation charter. Yet in spite of this, Engliš' unanimous election as rector stirred up great resentment among Communist Party representatives, which manifested itself in attacks on the university and on Engliš personally in the communist press.[73] Disputes with the aforementioned Minister of Information, Kopecký, who was behind part of the press campaign against Engliš and who, on his own initiative, invited a number of guests to the anniversary celebrations without the university's knowledge, dragged on from the autumn of 1947 until February of the following year.[74]

---

73 John Connelly, *Captive University. The Sovietization of East German, Czech and Polish Higher Education 1945–1956* (Chapel Hill: The University of North Carolina Press 2000): 114–118.

74 The short period when Engliš served as rector of Charles University during the anniversary year was recently described in detail, not only based on official documents, but also drawing on his unpublished memoirs and personal papers. Zdeněk Pousta, "Anglišův rektorský rok [Engliš' year as rector]", in: Petr Hruška (ed.), *Rok 1947. Česká literatura, kultura a společnost v období 1945–1948* (Prague: Ústav pro českou literaturu AV ČR 1998): 177–185; Jaroslav Čechura, "Rektorský rok Karla Engliše [Karel Engliš' year as Rector]", in: Petr Svobodný and Blanka Zilynská (eds.), *Věda v Československu v letech 1945–1953* (Prague: Karolinum 1999): 255–273; Blanka Zilynská, "Rezignace Karla Engliše na funkci rektora UK 26. února 1948 [Karel Engliš's resignation from the post of rector of the Charles Univeristy on 26 February 1948]", *Acta Universitatis Carolinae – Historia Universitatis Carolinae Pragensis* 39 (1999), no. 1–2: 125–127.

The University's Anniversary Overshadowed by the "Victorious February" of 1948

The final stage of the preparations for the jubilee was fundamentally influenced by a momentous political event: the takeover of power by the Communist Party of Czechoslovakia at the end of February 1948. Immediately after the war, the Communist Party was numerically the strongest, but not yet the dominant political party. However, the victory in the elections of 1946, subsequent participation in the government (including full control of the Ministry of Interior), and support by the Soviet Union, further strengthened their position. This enabled them to increase pressure on domestic opponents and, ultimately, even to cause a government crisis. On 20 February 1948, twelve ministers from different democratic parties submitted their resignation in the assumption that this would lead to the fall of the whole Communist dominated government. Unfortunately for them, it turned out differently. President Beneš was forced to accept the resignations a few days later and the departing ministers were replaced by politicians who collaborated with the Communist Party. The existing political regime broke down and adherents of the new one swiftly dealt with their opponents.[75]

The February takeover had far-reaching consequences, also for the university.[76] The purges that swept the whole of society also affected the teaching staff (including members of the anniversary committee) and the students, who were for the most part polarized in two irreconcilable camps. While the Communist supporters set up action committees in order to expel their resisting colleagues from the university, the democratically inclined students organized a march on Prague Castle. It was planned as the only open protest against the Communist Putsch, but even before it could take place, it was violently suppressed by security units. At the end of March, representatives of the Communist students took part in the last meetings of the committee for the anniversary celebrations.

The composition of the committee was also affected by the political purges. Hutter, director of the anniversary office and also in charge of the music for the celebrations, was not only expelled from the university but later even

75 On "Victorious February", see Karel Kaplan, *Der kurze Marsch. Kommunistische Machtübernahme in der Tschechoslowakei 1945–1948* (München: Oldenbourg 1981); Karel Kaplan, *Pět kapitol o únoru* [*Five chapters on February*] (Brno: Doplněk 1997); Václav Veber, *Osudové únorové dny 1948* [*The fateful days of February 1948*] (Prague: Nakladatelství Lidové noviny 2008).

76 Connelly, *Captive University* (2000): 127–132.

imprisoned.[77] And Engliš, as persona non grata, resigned just a few days after the Communist takeover, but surprisingly, his post was not filled by a Party faithful. This was probably due to lack of time, and in acknowledgement of the international reach of the anniversary. After a new election, the position of rector was filled by a non-political representative of the exact sciences, the previous rector Bohumil Bydžovský, who was known to have many contacts abroad.[78]

Although the replacement of the rector had a truly dramatic effect on the course of the jubilee, it was clear that the new regime could not afford to completely cancel the university anniversary and ignore possible international ramifications. Therefore, the celebrations took place on schedule, from 4 to 10 April 1948. The first day began with a ceremonial concert. During the following days, masses were held in various churches, exhibitions were opened (in addition to those mentioned above, there was also the exhibition "Artists for the University"), wreaths were laid on the tomb of the interwar President Tomáš Garrigue Masaryk, and participants were taken for excursions to Karlštejn Castle and to the historical town of Kutná Hora. The main international ceremony was held on 7 April in the Vladislaus Hall at Prague Castle. During the solemn meeting, the public observed President Beneš with sadness and mixed feelings. The gravely ill president, who for so many represented the continuity of the post-war republic with the interwar democratic Czechoslovakia, had withdrawn almost completely from public life after the events of February 1948. His speech at the university celebrations was his last public appearance before his death a couple of months later.[79]

The following day, honorary doctorates were awarded in the newly renovated great hall of the *Karolinum*. This ceremony in particular was significantly

77 In 1950, Hutter was arrested and two years later he was sentenced to imprisonment on trumped up charges, being accused of "high treason and espionage". He was released in 1956 by presidential amnesty and in 1990, he was fully rehabilitated. Tomislav Volek, "Profesor Josef Hutter – oběť dvou totalitních režimů [Professor Josef Hutter, victim of two totalitarian regimes]", *Hudební věda* 31 (1994), no. 4: 363–373.

78 Bydžovský was aware that thanks were due to Engliš. Once the celebrations faded away from public attention, he remembered him in a letter in which he expressed his warm gratitude and by issuing a commemorative medallion. This human approach contrasted sharply with the growing political persecution of the former rector. Institute of the History of Charles University and the Archives of Charles University, collection Akademický senát Univerzity Karlovy, box no. 62, inventory no. 851: *Letter from B. Bydžovský to K. Engliš* (18/09/1948).

79 See Zbyněk Zeman and Antonín Klimek, *The Life of Edvard Beneš 1884–1948. Czechoslovakia in Peace and War* (Oxford: Clarendon Press 1997).

affected by the boycott of Western, especially British and American, representatives. Almost without exception they refused to accept the honorary degrees offered to them, so that mainly scholars from Eastern and Central European universities were honoured. From the 74 universities, which had originally confirmed their participation, in the end only 46 were represented. The official written refusals to join in the celebrations in general and in the ceremony of acceptance of honorary doctorates in particular, mostly stated a disagreement with the political situation and the recent changes at the university (see figure 2.6).

This was voiced in the most succinct and yet most detailed way by a representative from the University of Liverpool:

> We now announce with the greatest regret that we withdraw our acceptance of that invitation and refrain from sending our representative. The recent dismissal of professors and their replacement by others have been carried out under conditions which violate the historic autonomy of the University and deny the essential principles of academic freedom.

FIGURE 2.6 *The award of honorary doctorates in a solemn ceremony on 8 April 1948 in the great hall of the* Karolinum, *with a view on the new statue of Charles IV and the tapestry with the motif of the oldest university seal and symbols of the four original faculties.*
PICTURE COURTESY OF THE INSTITUTE OF THE HISTORY OF CHARLES UNIVERSITY AND THE ARCHIVES OF CHARLES UNIVERSITY

> Our acceptance of the invitation to an academic function has already been misrepresented as support for a political party. We protest against that misrepresentation and against the action which has undermined the liberties of the University; and we cannot appear to condone that action by our participation in the celebrations. The Charles IV University has had a long and honourable history; we deplore the unmerited misfortune which has now befallen it; and we look forward to the day when it will regain its freedom and will resume its place in the world of learning.[80]

Despite the non-participation of Western universities, the celebrations of the Prague University's anniversary did not pass without response from abroad. Even the prestigious magazine *Nature* printed a relatively detailed report on it.[81] Characteristically, its author, John Gerald Frederick Druce, a British chemist and one of the discoverers of the element rhenium, was a long-time supporter of Czechoslovakia and had published several books on this Central European state since the early 1930s. He almost certainly participated in the jubilee as a private individual, and not as the official representative of his university. He had been invited on the initiative of his former fellow student Jaroslav Heyrovský of the faculty of natural sciences, a physical chemist and the only Czech scientist to be awarded the Nobel Prize.

A View from Inside: The Jubilee as Depicted in the Diary of One of the Organizers

Also from within the university, some criticism was expressed on the recent course of the preparations and the jubilee itself, yet never in public of course. The inside view of the historian Otakar Odložilík in his recently published diaries is revealing in this respect. Like many other lecturers who were overwhelmed by the exceptional post-war conditions, his involvement in the anniversary preparations was not necessarily a welcome additional duty. Odložilík was a specialist in early Czech history and one of the first authors of the anniversary publications, which had already been planned in the mid 1930s. He spent the war in exile in the United States and Great Britain, where as

80 Institute of the History of Charles University and the Archives of Charles University, collection Akademický senát Univerzity Karlovy, box no. 63, inventory no. 857: *Letter from the University of Liverpool to the rector of Charles University* (25/03/1948).

81 Gerald Druce, "University of Prague Sexcentenary Celebrations", *Nature* 161 (1948), no. 4096: 670–672.

early as 1944 he secured in Oxford a promise of participation of English universities at the anniversary celebrations.[82]

Comments regarding the anniversary appear prominently in the pages of his diaries until the autumn of 1947, when he wrote an eighty-page outline of the university's history.[83] Odložilík at first refused to write this book. He gave in only after repeated insistence of Rector Engliš, at the point when it was clear that the planned collective work on the past and present of the university would not be ready in time for the celebrations. He had about two months (October and November 1947) to write the text. In his diary, he noted on 23 November 1947:

> But I also had to take on a commitment assigned to me by Rector Engliš and set about writing a book on the university. I did it by not lecturing last week, thus gaining a little bit more time around last and this Sunday. I started last Friday and have been at work almost non-stop since then [...]. I don't have a finished manuscript yet, but I have come to the second last page. What I have written are seven little chapters and a two-page introduction. Overall it will be somewhat more than fifty pages of typewritten text. It has so tired me out that I want nothing else than to get ready with this obligation of writing an order at such a pace.[84]

The brochure was supposed to be ready by 6 April, together with its English, French, and Russian translations. Soon after the Communist takeover, however, even Odložilík's work as university historian was affected by the interference of Communist students. At the end of March he noted:

> The students have stopped the printing of the publication on the university and I wouldn't have known about it [...]. It is not very likely that the celebrations will take place. Two of them visited me on Thursday afternoon and we agreed on the final formulation of some passages of which they had objections against the wording.

82 Otakar Odložilík, *Deníky z let 1924–1948. Vol II. (1939–1948)* [*Diaries 1924–1948. Vol. II (1939–1948)*], edited by Milada Sekyrková (Prague: Výzkumné centrum pro dějiny vědy 2003): 454, 496.

83 Otakar Odložilík, *Karlova Universita 1348–1948* [*Charles University 1348–1948*] (Prague: Karlova Universita – Orbis 1948).

84 Odložilík, *Deníky* (2003): 576–577.

Odložilík came back to this incident in a note written during the celebrations (on 6 April 1948):[85]

> My book came out, but so few copies were printed that it could not be distributed as originally intended. I just glanced over a copy in a hurry this afternoon and got the impression that they prepared it for publication very nicely. If it hadn't been for the to-do with the students, it would all have been just fine, even though we started late.

This was clearly a situation unimaginable under normal circumstances. It should be added that later that year, Odložilík once again left to American exile.

The Interrupted Jubilees of 1848 and 1948: Parallels and Specific Features

One of the key characteristics shared by both disrupted celebrations was the spontaneous activity of the students who found a way of expressing their radical views despite the unforeseen and dramatic change of circumstances. Yet in both cases, their attitude was such that no consolidated regime would permit it on a large scale.[86]

On the other hand, the two jubilees differed largely in their attitude to the national question. The 1848 jubilee was in this respect completely neutral. Preference for the German language was given out of consideration that it was the official language of the empire. But like the records of the preparatory committee show, the organizers tried, within reason, to maintain a certain balance between the two languages of the land. Even so, the jubilee was at least to a certain degree used as a political demonstration in the course of the student celebration on 7 April 1848. In the case of the 1948 celebrations, there were clear efforts to use the anniversary largely for nationalist and political purposes. Taking into account post-war anti-German sentiments, the organizers

85 Odložilík, *Deníky* (2003): 582–583.

86 One such example are the celebrations of the 600th anniversary of the foundation of the Jagiellonian University in Krakow in 1964, where interventions of the state authorities left the university and its students only rather limited space to do their own planning and rehearse their role in the jubilee. Instead, the regime tried to use the anniversary for its legitimisation and self-promotion. Urszula Perkowska, *Jubileusze Uniwersytetu Jagiellońskiego* [*Anniversaries of the Jagiellonian University*] (Kraków: Wydawnictwo i Drukarnia "Secesja" 2000): 205–206.

of the celebrations at first completely suppressed anything that would deal with the link between the university and the German language or community. After the coup d'état, the Communists then excluded from the celebrations most of their actual and potential political opponents. They tried to adjust the tone of the celebrations according to their ideological convictions and political orientation towards the Soviet Union and the states of the Eastern Bloc.

Despite the detrimental impact that unexpected changes in the political situation had on the largely failed jubilee celebrations, the efforts also brought some lasting positive results. Several valuable publications came out, the university obtained funds for an ambitious reconstruction of its oldest buildings, and several valuable works of art improved their interior. The statue of Charles IV and the renovation of the *Karolinum*'s Aula Magna became important and well-known symbols of the academic tradition in Prague. So although neither of the university celebrations proceeded according to plan, they are both associated with concrete results of lasting value. In both cases, the jubilee also left a clear mark on the face of the Czech metropolis: in 1848, one of the most prominent memorials in Prague was created and, despite all the turmoil of the time, unveiled, and in 1948, the oldest university building in the very heart of Prague's Old Town had undergone the reconstruction that had been debated for over a century.

CHAPTER 3

The Royal Frederik University in Kristiania in 1911

Intellectual Beacon of the North – or "North Germanic" Provincial University?[1]

Jorunn Sem Fure

During the jubilee, two dilemmas that had already lain in slumber for a long time were brought into the public domain. The first of them concerned the task of the university. Was it to be in the first place an institution for the education of the people (et folkeuniversitet), *as Bredo Morgenstierne promoted in his extensive history of the university, published for the occasion – or rather a research institution, as the then rector, Waldemar Christopher Brøgger, defended in his speeches and in a pamphlet with an obvious historical slant to it? Secondly, the university itself, as well as the foreign guests who attended the jubilee, perceived the university (and its history) both as a crucial institution for building up the nation, as well as having contributed extensively to the international scientific community. Although at the time of the jubilee, the elements of both dilemmas were mainly discussed as opposing ideals, looking at them in retrospect, the developments during the interwar years clearly show that the university was, to some extent, able to realise both of them in unison. Therefore, the Royal Frederik University in Kristiania, rather than being a "North Germanic" provincial university, can be depicted in 1911 as an intellectual beacon of the North.*

It was early September in 1911, the main street in Kristiania, Karl Johans Gate, leading up to the university and the Royal Palace, had taken on an unusual appearance in the mild autumn weather. Norwegian flags, green garlands and red banners adorned the pavements and, in the evenings, multi-coloured lanterns gave the university buildings and the surrounding grounds and gardens a touch of magic. Groups of foreign-speaking gentlemen walked along the street, many of them accompanied by their wives and daughters. Those gentlemen were distinguished academics from every part of the world, wearing red, gold

1 The present chapter is based on the text of my inaugural lecture as Heinrich Steffens Professor at the *Humboldt-Universität zu Berlin*, 2 February 2010. Adriana Alexander has translated the original text from German into English and also edited the final English version. Professor John Peter Collett has given a substantial contribution by suggestions and comments to the text in several stages.

 | DOI 10.1163/9789004265073_004

FIGURE 3.1 *Atmospheric picture of the main street, Karl Johans Gate, during the jubilee. Photographer: A/S Abel.*
PICTURE COURTESY OF THE MUSEUM OF UNIVERSITY HISTORY, UNIVERSITY OF OSLO

or black gowns of silk and velvet, some of them with golden chains around their necks and square caps upon their heads. There were rectors of German universities and deans of American faculties, and their colourful attire was quite different from the sober black and grey morning coats of the Norwegian professors and deans (see figure 3.1).

Between the second and the tenth of September 1911, the Royal Frederik University in Kristiania had invited many dignitaries to the celebration in the capital, and alumni living all over Norway were encouraged to don their student caps and organize events in their localities. Every single university and scientific institution from around the world was invited to celebrate its first centenary with a series of large-scale events.[2] Public figures and

2 The programme, the speeches, the congratulatory telegrams and the guest lists were published in Frederik B. Wallem (ed.), *Det kongelige Frederiks Universitets Hundreaarsjubilæum 1911. Festberetning utarbeidet efter opdrag av det akademiske kollegium* [*The centenary celebration of the Kongelige Frederiks Universitet in 1911. Commemorative book preprared under the authority of the academic council*] (Kristiania: A.W. Brøgger Bogtrykkeri 1913). Unless otherwise stated, the references are to the speeches contained in this album. See Jorunn Sem Fure,

representatives of academia had been invited to send delegates to Kristiania for the occasion, and the organizers had hoped for as many as 300 to 400 guests from abroad. In the end, the number of foreign delegates who actually did arrive in Kristiania at the beginning of September 1911 was 140, not including accompanying spouses and other family members. Still, this was quite impressive. No international event of this scale had ever taken place before in Kristiania.

Was a celebration on such a scale the expression of Norway's confidence after its consolidation as an independent state and the country's pride in the achievements of its only university? Or was it intended to free them both from a perceived provincialism?

By inviting guests from foreign universities, the hosting Norwegian university clearly wanted to present itself as part of a wider international community and, thus, take its place in the long university tradition, dating back to the foundation of the University of Bologna eight hundred years earlier. However, there was little mention during the week of celebration of the actual historic roots of the Kristiania University. Of course, as it is customary on such occasions, some publications were released in which the history of the university passed the revue, but, as will become clear, these writings served the goals of their authors at the time, rather than meeting historiographic demands. This was in striking contrast to the University of Berlin, which had celebrated its centenary the year before in 1910. On that occasion, a series of texts written by members of its first generation of academic staff had been reprinted, since they formed part of that university's heritage – among them works by Fichte, Schleiermacher, Steffens and Humboldt. However, at the Norwegian jubilee no specific reference was made to the University of Berlin, or to the concept which later became known as the 'Humboldt university model'.[3]

In light of the prominence accorded to the 'Humboldt model' in later writings on the modern European university institution, the absence of Humboldt altogether in Kristiania in 1911 may be surprising. However, as the Norwegian historian Sivert Langholm has shown, it was not until well into the 1960s that references to the 'Humboldt model' or the 'Humboldt university

"2. Universitetets hundreårsfeiring", in: Idem, *1911–1940. Inn i forskningsalderen* (Universitetet i Oslo 1811–2011 3) (Oslo: Unipub 2011): 31–55.

3 See pp. 183–207.

ideas' were introduced into Norwegian academic writing on the nature and functions of a university.[4]

Nevertheless, every Norwegian academic in 1911 would have been well aware of the great importance of contemporary German universities on the Norwegian university and for science and humanities in Norway as a whole. In fact, a large number of Kristiania University teachers had been trained in Germany. Having obtained a first degree in Kristiania, the majority of Norwegian-born scholars were required to spend time abroad before their career could culminate in a professorship at home. 'Abroad' would mostly mean Germany (including German-speaking Austria) – or, to a lesser extent, France. Aspiring scientists, historians and philologists were by no means the only Norwegians to go abroad for their education. German universities, *technische Hochschulen* and academies of art were the institutions in which Norwegian engineers, architects and artists received their training, because Norway still lacked specialized educational institutions at this level.

During the university jubilee, however, there was not much mention of the extent to which Norwegian science and academic culture depended upon Norwegian scholars' years of study abroad. Generally, it was not a time to talk about foreign historic models of higher education, but instead to concentrate on national themes, whether historical or contemporary, as well as on visions of the university's future – and the nation's.

The university's leadership wanted the university to be seen as a largely autonomous cultural institution, playing a central role in the life of the country. An extensive history of the university, written by the professor of law Bredo Henrik von Munthe af Morgenstierne and published for the occasion,[5] emphasized the university's contributions to the development of Norway's state institutions, her economy and her culture since the university was founded in 1811. His book also highlighted the university's wider role in the education of the Norwegian people.

The university sought to convey its historical and contemporary importance as a teaching institution serving the country and, at the same time, to

---

4 Sivert Langholm, "Das 'Humboldt-Model' in Norwegen. Symbol, Begriff und Wirklichkeit", in: Rainer Christoph Schwinges (ed.) *Humboldt International. Der Export des deutschen Universitätsmodels im 19. und 20. Jahrhundert* (Veröffentlichungen der Gesellschaft für Universitäts- und Wissenschaftsgeschichte 3) (Basel: Schwabe Verlag 2001): 205–230.

5 Bredo Henrik von Munthe af Morgenstierne, *Det kongelige Fredriks universitet 1811–1911* (Kristiania: Aschehoug 1911).

demonstrate to the international academic community that it was also a scientific research institution able to represent Norway on an equal footing with its foreign counterparts. The jubilee speeches praised the achievements of Norwegian scholars at an international level and pointed to those they might yet make in the future. The jubilee thus became an opportunity to demonstrate the university's achievements to the public both at home and abroad.

The celebrations started on 2 September with the inauguration of the impressive new Aula, followed by a ceremony for the matriculation of new students. On Monday 4 September, the foreign guests attended a banquet in their honour and, on the following day, guests from home and abroad handed over their letters of congratulation at a lengthy ceremony.[6] The Rector, professor of geology Waldemar Christopher Brøgger, delivered the main jubilee speech and words of thanks (in German). On Wednesday 6 September, no less than 93 honorary doctorates were conferred on foreign academics at another lengthy ceremony in the Aula, before the day culminated in a banquet held at the Royal Palace, with the King and Queen as hosts. On Thursday, the guests were taken on tours to visit scientific collections, museums, the Botanic Garden and other university institutions.[7] Two public lectures that day attracted much attention: the Kristiania professor of physics Kristian Birkeland spoke (in French) about the electric and magnetic forces of the solar system,[8] while one of the prominent German guests, Karl Lamprecht, professor of history in Leipzig, held a lecture (in German) drawing parallels between the cultural history of the German and Norwegian peoples (see figure 3.2).[9]

6 The congratulatory addresses – over 240 – were either handed over by the guests attending the celebrations, or sent by mail. The letters were beautifully crafted, most of them using medieval calligraphic models, with the great seals of the senders attached. Several were presented in boxes inlaid with gold, precious wood, velvet, and leather. In 1912, the letters were displayed in the Kristiania Museum of Applied Arts (*Kunstindustrimuseet*) for the public to admire. At the second centenary celebrations in 2011 they were again brought out of their boxes and capsules and exhibited in the University Library in Oslo.

7 In 1911, new buildings for the Archeological Museum and Ethnography Museum, and the Botanic Museum, were just taken into use. A range of other building projects was in the planning stage. During the following ten years the Museum of Geology and the Museum of Zoology were added, as well as a new University Library.

8 For a biography of Birkeland, see Lucy Jago, *The Northern Lights. The true story of the man who unlocked the secrets of the aurora borealis* (London: Hamish Hamilton 2001).

9 This lecture was reviewed in the *Leipziger Tageblatt* (19/09/1911). Lamprecht was very influential in Norway as a teacher and a role model for Professor Halvdan Koht and other Norwegian historians who had studied in Germany.

FIGURE 3.2A
The programme of the jubilee included academic as well as social events and aimed to promote the university as a national and international cultural institution.
PICTURE COURTESY OF THE NATIONAL ARCHIVES OF NORWAY (RA/S-2868)

After a whole week of festive events, guests and hosts boarded on the train on Friday morning for a visit to Bergen, admiring breathtaking views of the Hardangervidda mountain plateau on the way. Having toured the scientific institutions of Norway's second largest city, the party returned by train to Kristiania, arriving in the evening of Sunday 10 September. With this, the jubilee was concluded and the university's next century had begun.

SØNDAG 3. SEPTEMBER:

Kl. 11 form. i Vor Frelsers Kirke:
Festgudstjeneste med præken av det Teologiske Fakultets senior, professor Michelet.
Sang av Studentersangforeningen.

MANDAG 4. SEPTEMBER:

Kl. 5 eftm. i Universitetets gamle festsal (domus academica): Orienteringsmøte for de delegerte.

Kl. 8 aften i Grand Hotel: Souper git av Universitetets lærere, doctores m. fl. for de utenlandske delegerte med damer.

Antræk: Selskapsdragt.

TIRSDAG 5. SEPTEMBER:

Kl. 11 form. i Nationalteatret:

1. Universitetskantaten „Lyset" av Bjørnstjerne Bjørnson og Otto Winter-Hjelm. 1ste – 3dje avdeling. – Studentersangforeningen og Nationalteatrets orkester under ledelse av O. A. Grøndahl.
2. Tale av rektor.
3. Taler av repræsentanter for de delegerte og avlevering av adresser.
4. Kantatens 4de avdeling.

Antræk: Gala – embedsdragt – uniform.

Kl. 8 aften: Souper paa det Kgl. Slot.

Antræk: Civil gala.

FIGURE 3.2B (cont.)

A People's University or a Research University?

What image did the university leadership want to project vis-à-vis the foreign guests and the Norwegian public? What impression did the jubilee celebrations and the university itself make on the attendees from abroad? And how did the Norwegian public see its university in those days?

The jubilee speeches and publications reflected two views of the university that did not necessarily contradict each other, but could actually be seen as complementary. In his book on the history of the university, Bredo Morgenstierne,[10] who was to succeed Brøgger as rector a few years later, described the changes that the university had undergone since the middle of the nineteenth century. During this period, scientific research had gained ground and, simultaneously, academic freedom had allowed a new kind of autonomy and structuring of scientific fields within the university. The university had experienced the triumph of Realism over Classicism and the rise in status of the Norwegian language: "The majority of the population saw an education based on the classics as elitist, as something which was alien to them and which was not compatible with contemporary ideals of democracy and national consciousness."[11]

In Morgenstierne's work, the university appears partly as a scientific powerhouse and partly as an institution for the education of the people: *et folkeuniversitet* – a people's university. He stressed that recruitment should be made from a wide range of the population, hailing the entry of women into academic life and science as essential to this. The first female student was admitted in 1884, and in 1903 the first woman was awarded a doctoral degree.[12] The first female professor was appointed in the year following the jubilee, Kristine Bonnevie, professor of zoology from 1912.

Morgenstierne regarded the university as an institution of higher education which had constantly integrated new social groups and would continue doing so. The knowledge accumulated in the university would benefit ever-larger parts of the people and, in this way, both the social importance and the influence of the university would maintain its growth. These qualities would make the university an institution working for the people, run by the people.

According to Morgenstierne and those who shared his ideas, the main characteristics of "a people's university" were:

- admission of students from all social backgrounds;
- encouragement of students to see themselves fundamentally as carriers of knowledge and skills that would benefit the whole population and not only urbanised elites;
- selection of teaching staff more on the basis of their ability and experience as educators, rather than of their scientific qualifications;

10 Morgenstierne, *Det kongelige Fredriksuniversitet 1811–1911* (1911).

11 Morgenstierne, *Det kongelige Fredriks universitet 1811–1911* (1911): 312.

12 Ernst Haakon Jahr, *Clara Holst. Kvinnelig pioner i akademia* (Oslo: Novus 2006).

- according the highest priority to education and training, ranking these tasks higher than the university's relations with the international scientific community;
- encouragement of university staff to see their main task as making their knowledge available and accessible to a wide public, which would then, in turn, respond with respect and appreciation.

In contrast to this, a different, research-oriented perspective was put forward by two prominent speakers at the jubilee: Fredrik Stang, professor of law at the Royal Frederik University, and the university's rector himself, Brøgger. In their eyes, a university was a transnational institution linked to a family of universities worldwide, but also of crucial importance to its home nation. To them, pushing back the frontiers of knowledge and thereby achieving progress for mankind were the true attributes of science, disregarding any national borders. However, in order to maintain Norway's position on a par with the most developed countries, it was, in their view, of pivotal importance that the country strengthened the potential of its research capabilities. A nation that could not keep pace with the international scientific competition would face an uncertain future.

In Stang's speech to the foreign honorary doctors, held on 6 September 1911, he stated:

> Those men struggling on the edge of human knowledge, where light ends and darkness begins, where day is swallowed by a looming, perilous night all around us, those men are precious to their people because every tiny glimmer of light they set free is a great step forward for human life.[13]

Who were those men, and what were the characteristics of their work? Just as the art of warfare had changed over the centuries, said Stang, so had science. Combat between single knights had given way to massive military deployment with structured leadership and costly equipment. Similarly, there were now few scholars achieving breakthroughs as individuals. Like the art of war, present-day science demanded planning, leadership, and teamwork.

In his speech, Stang conveyed a modern concept of research, inspired especially by developments in the larger countries in Europe and in the United States, where scholars worked in organized collectives, in laboratories, in seminars and

13 Wallem (ed.) *Det kongelige Frederiks Universitets Hundreaarsjubilaeum 1911. Festberetning* (1913): 142.

institutes, and addressed their tasks collectively. Stang admitted that this was a well-established reality abroad, but in Norway still a far-off dream, thereby pouring cold water on his fellow Norwegians' pride for their university.

The university in Kristiania had never progressed beyond its initial phase, Stang concluded, and it was ill equipped to take up the challenges of the new reality: "We cannot employ as many scientists as we need – every academic is mostly on his own, without colleagues or support staff."[14]

The alternative to a people's university that Stang had introduced in his speech was that of "a research university", a model that was further elaborated upon by Brøgger in his later lengthy speeches at the jubilee. The main criteria of a research university, for Stang and Brøgger, were the following:

- Scientific talent and achievement would be seen as pre-requisites for academic appointments, ranking just as highly or higher than teaching ability and experience.
- Teaching should be based as far as possible on research – students should acquire scientific reasoning and be introduced to conducting research themselves.
- Professors should play a public role different from their nineteenth-century predecessors. Their authority should no longer rest on their erudition and general judgment, but on their role as active scientists, as experts and specialists in their respective fields of research, and on the originality of their discoveries.
- Scientific practice should be given a new basis. Philosophical observation with an *a priori* bias would be replaced by a commitment to empirical research aimed at making new discoveries. The laboratory would be the scientists' typical work place, the microscope his working tool – and the two would represent to the outside world the new face of science.[15]
- Research should not be seen as a desirable activity alongside teaching, but would be built into a university's everyday life, permeating its budgets, building designs and organizational structures. Participation by the faculty

14 Wallem (ed.) *Det kongelige Frederiks Universitets Hundreaarsjubilaeum 1911. Festberetning* (1913): 141.

15 For further analysis of the concept of a research university in Norway in 1911, see Jorunn Sem Fure, "Forskningsuniversitet – Retorisk ideal eller realitet [Research university – A rhetorical ideal or a reality]", in: John Peter Collett, Jan Eivind Myhre and Jon Skele (eds.) *Kunnskapens betingelser. Festskrift til Edgeir Benum* (Oslo: Vidarforlaget 2009): 193–218.

in international networks and organizations, publishing and conferences would be expected and encouraged appropriately.

Morgenstierne and Stang advocated two different views of what the main purpose of a university ideally should be – "the people's university" versus "the research university". Nevertheless, they both considered it possible and desirable to combine these two ideals within one institution.

Brøgger, for his part, had been expressing criticism of the university in Kristiania ever since his return to his alma mater in 1890, after nine years as professor at *Stockholms Högskola* in Sweden. In his view, the Kristiania University was little more than a school for civil servants where scientific activities had stagnated.[16] The academic staff was burdened with too many teaching and examining duties and tasks to have time, money and space for research. More than any of his contemporary university colleagues, Brøgger actively sought other models and new organizational solutions that would allow research and scientific development to grow in Norway. In the years from 1900, he had put forward a series of initiatives, proposals and ideas. At one time, he proposed creating a separate Academy where a small group of Norway's most distinguished scientific talents could work, exempt of teaching or administrative commitments. As members of this Academy, they would be dedicated exclusively to their respective branches of science and humanities.

Brøgger's main concern was to free research from the slavery of duties incumbent on academics because of their status as the teachers of civil servants, which had been the traditional function of the Norwegian university professors of the nineteenth century. They had devoted their lives chiefly to the training of priests, jurists, doctors and schoolteachers. Instead of the traditional university model, Brøgger envisaged the creation of some 20 extraordinary professorships attached to the Academy, which was to be granted a large degree of autonomy vis-à-vis the state. The models for the Academy that he had in mind were the German *Kaiser-Wilhelm-Gesellschaft zur Förderung von Wissenschaft und Forschung* (KWG), which had been founded as an independent research organization in 1911, and also *Stockholms Högskola*, which was founded as a privately funded alternative to the traditional Swedish state universities of Uppsala and Lund.

In his speeches at the jubilee, Brøgger went further (see figure 3.3). Instead of merely proposing a national Academy, he addressed the foreign guests with

16 For more on Brøgger's life, see Geir Hestmark, *Vitenskap og nasjon: Waldemar Christopher Brøgger 1851–1905* (Oslo: Aschehoug 1999).

FIGURE 3.3 *On 2 September 1911, Rector Brøgger opened the centenary celebration with a speech in which he pled for a rejuvenation of the academic staff at the university.* Aftenposten (*02/09/1911*).
PICTURE COURTESY OF THE NATIONAL ARCHIVES OF NORWAY (RA/S-2868)

a proposal for no less than a world research centre, in which the world's greatest minds, its geniuses, would gather together and dedicate themselves solely to the development of their talents and the cultivation of science.

Brøgger regarded science and scientific research as an activity for geniuses, for outstanding people with uncommon talents. In his view, such men were motivated not by personal ambition or the desire for individual advancement, but merely by an unselfish will to serve their countries and their peoples. International recognition for their work would confer higher status not only on them, but also on their countries of origin. Consequently, it was in the interest of all countries that such people be freed from problems with funding and other mundane worries.

As rector from 1907, Brøgger wanted to raise the Kristiania University's scientific prestige and at the same time encourage a constant exchange between the university and scientific institutions outside the university, such as the one that was foreseen in Berlin between Berlin University and the new institutions put up by the KWG. Consequently, Brøgger was confronted with a rhetorical dilemma at the jubilee: on the one hand it was important for him to emphasize the benefits and triumphs of science in order to win enthusiastic support for further research, whilst on the other hand, he had to describe conditions for research in Norway at their worst and most deprived, in order to argue for a necessary political commitment to substantial growth in university budgets.

The picture Brøgger painted of his own university at the 1911 jubilee was remarkably distressing. The university did not have enough money, nor functional buildings or enough qualified support staff. Many of the professors were too old to retain what Brøgger called their "intellectual elasticity". There were no fixed pension arrangements, so a professor could simply not afford to retire. 20% of the professorial staff was over 70, some even in their eighties. Just as his opponent Morgenstierne had done, Brøgger used the history of the university to enforce his argumentation to some extent. He published a manifesto pamphlet to attract broad public support, *Vort universitetet. Dets midler og dets maal* [*Our university. Its means and its goals*]. In this call-to-arms, he portrayed what he considered a troubling situation and weaved into this account why the nation should support academic research. He reminded readers that the university must serve not only as an educational institution but also one devoted to research. Given the lack of other similar institutions, the university must assume the role as "the centre for the country's entire scientific life". Using the language of a hundred-year-long "struggle for survival", Brøgger called upon the whole populace to join, as an obligation, in supporting the university as an institution indispensable for the entire nation.[17]

17 Waldemar Christopher Brøgger, *Vort universitet. Dets Midler og dets Maal* [*Our university. Its means and its goals*] (Oslo: A.W. Brøgger Bogtrykkeri 1911). See Robert Marc Friedman, "Science, Populist Democracy and Honour. The 1911 Centenary Celebration of the Royal Frederick University in Kristiania", in: Pieter Dhondt (ed.), *National, Nordic or European? Nineteenth-Century University Jubilees and Nordic Cooperation* (History of Science and Medicine Library 25. Scientific and Learned Cultures and Their Institutions 4) (Leiden: Brill 2011): 280–281.

Norway had fostered great minds, Brøgger insisted, pointing to the great names in the nation's scientific hall of fame, such as the mathematicians Niels Henrik Abel and Sophus Lie. Among his contemporaries, Brøgger particularly mentioned the pioneers of oceanography, an emerging discipline, in which Johan Hjort, Fridtjof Nansen and Bjørn Helland Hansen were internationally renowned. Kristian Birkeland, though his new theories about the *aurora borealis* remained controversial, had also become world-famous.

Through every argument that Brøgger and Stang put forward, they defined the university as first and foremost an institution for research, and the Kristiania University, in particular, as an institution needing immediate support in order to keep up with this ideal. A stagnating university would never be able to compete with leading research institutions in the rest of the world. In the present stage of scientific development, and with fierce international competition between researchers, there were only two alternatives: expansion or extinction. A university, or indeed a whole nation that was not able to keep up in the scientific race, would be doomed to decline, and, in the longer term, their survival would be threatened. The research university, therefore, needed constant growth, in budgets, in manpower and in facilities, and this was of pivotal importance to the whole nation's future.

The people's university, dedicated to education, growing hand-in-hand with the nation's cultural development, would also need to be improved and constantly renewed, but such a university did not depend on spectacular breakthroughs in knowledge, or scientific excellence achieved by a small number of people of extraordinary talent. Such a university aimed at conveying knowledge and extending opportunities to a large number of people, serving the curiosity and desire to learn of a growing number of students, and thereby benefitting the nation at large.

The University Seen from Outside

> What achievements, what intellectual power, how much that is praiseworthy – all coming together in this new century [...]. May the torch borne by the university in Christiania shine just as bright throughout the centuries![18]

The congratulatory letter from which this quotation is taken was signed by the rector of the Imperial and Royal Franz Joseph University in Czernowitz, a province in the then Austro-Hungarian monarchy. The phrasing is representative of

18 This and the following quotation are contained in Wallem (ed.) *Det kongelige Frederiks Universitets Hundreaarsjubilaeum 1911. Festberetning* (1913).

the almost 200 congratulatory messages – most of them written in Latin – which were received by the Kristiania University. All the way from the banks of the Volga, the academic senate of Saratov, the newest university in the Russian Empire, sent a letter calling the university in Kristiania "the beacon of science in the far North", while the university in Greifswald praised its sister institution in Kristiania as "the intellectual centre of Norway". The Norwegian university could bask in the sunshine of such flattering epithets, although they probably had been taken from the standard repertoire of phrases suitable for such occasions.

Disregarding such flourishes, what image did the foreign attendees at the jubilee actually have of Norway's one and only university? Those who actually had taken the trouble to check the historical facts and the origins of the Kristiania University seemed to have a very clear picture of it. Their greetings emphasized the importance of the foundation of the university as a turning point in Norwegian history and the emergence of the modern Norwegian nation. The successful campaigning for the university had strengthened the self-awareness of the Norwegian people and prepared the secession from the union with Denmark in 1814. The context of its foundation had deeply influenced the shape and further development of the Norwegian university. The university had been a core institution for the country. The emphasis put on the cultivation of Norway's own history, language and cultural heritage – as well as on contributing to the material development of the country – had overshadowed general European ideals of education. The Kristiania University had long ago bid farewell to the classics tradition, so firmly embedded in the European university elsewhere. Appreciating this important role of the Kristiania University for Norwegian nation building, the congratulatory addresses also recognized those Kristiania scholars whose contributions had a resonance beyond the country's borders. The Norwegians may have broken away from the international Latin-language academic tradition, but had been present in the international community of science and humanities.

Reporting from Kristiania, the German press noted, with some embarrassment, that the German delegates to the jubilee festivities had behaved in such a condescending manner as if they regarded the Norwegian university as a "North Germanic provincial university" (see figure 3.4).[19] The Norwegian daily newspaper *Aftenposten*, for its part, criticized the university's leadership for having chosen German instead of English as the main language of the jubilee. English was a world language, the paper insisted, and the largest delegation of guests actually came from the USA, the second largest from Britain. A German

19 The home and foreign press reports about the jubilee are gathered in the university records kept at the National Archives in Oslo.

FIGURE 3.4 *Academic procession headed by Rector Brøgger after the ceremony of awarding honory doctorates, as depicted in the German newspaper* Nationale Zeitung *(1911), no. 38.*
PICTURE COURTESY OF THE NATIONAL ARCHIVES OF NORWAY (RA/S-2868)

journalist also mentioned some friction arising out of the choice of language and thought everything could have been simpler: in other countries everybody could have used Latin, but because Norway had cast away the classics, there was now probably only one professor in the whole country who could speak Latin properly: the well-known linguist Alf Torp!

A journalist reporting for the *Manchester Guardian* made a detailed analysis of the social ranking and the relationship between the university and its environment. He found the ceremony for the matriculation of new students very impressive, with the King awarding diplomas and prizes to outstanding students. During the banquet held at the palace, His Majesty was referred to in an almost familiar tone as "the neighbour across the park".[20]

20 Charles Harold Herford, *Manchester Guardian* (19/09/1911). Herford was professor of English literature at the University of Manchester, known for his critical voice. "His weapon was always his pen, and his letters and articles in the Manchester Guardian and other journals showed how deeply he was stirred by whatever seemed to him social injustice and political tyranny. He never regarded scholarship as an end in itself; it was the implement of criticism with which to interpret the speculations of master minds," as it is noted in his obituary in *The Times* (21/04/1931), no. 45807: 17.

The proximity between the Royal family, the parliament and the university, not simply in terms of where they were located, but also socially and politically, struck that British journalist as an enviable characteristic of the egalitarian Norwegian society. But just how egalitarian was it? In the Norwegian press there were voices criticizing the university's leadership because the jubilee celebrations had not really involved the people. The city's inhabitants had just stood by as onlookers, watching the academic celebrities and the royal couple arriving and leaving.

There had been no people's celebration such as the one in Berlin in 1910, with students parading in a carnival atmosphere, free beer, traditional costumes, music and thousands of townspeople in the streets. There had been nothing like that in Kristiania – apart from a torch procession and a garden party with sausage on the menu. There had been little pageantry surrounding the Norwegian academics and students. As was the Kristiania tradition, neither professors nor students wore gowns. From its start, the Kristiania University had been sceptical towards the display of academic status. What there had been of academic pomp, was been abolished in 1845. The reasons for this would be dismissed by some as a historical misunderstanding. The rituals played out in costumes, neck-chains, rings, hats and other regalia were interpreted in Norway as representing aristocratic, elitist and monarchist values which were regarded as alien to the nation, and not as an expression of academic dignity, autonomy and authority.

All in all, it would be fair to say that the jubilee had demonstrated to the foreign guests that the university was small and belonged to a peripheral country, but was still imbued with vitality and ambition. The close relationship between the academics and students at the university, and the Norwegian state, the nation and the people, was another trait that struck observers from outside. The university was an important institution for all, and sobriety and seriousness had characterized the celebrations. An American professor even remarked that he would have liked to have taught Norwegians how to relate to their alma mater with more passion, love and enthusiasm.[21]

Is Scientific Practice Country-Specific or Universal?

In the written and oral addresses, both the Norwegians and their guests declared that science stood above national and political borders. Particular mention was made of the discipline in which Norwegian scientists had attained

21 Gisle Bothne (delegate from Minnesota), "Speech at the Norwegian-American breakfast", in: Wallem (ed.), *Det Kongelige Frederiks Universitets Hundreaarsjubilæum 1911. Festberetning* (1913): 13.

the highest renown worldwide: mathematics, the field which perhaps can be said to depend least on its cultural context.

A series of distinguished names was mentioned, personifying the scientific achievements linked to the University of Kristiania. Unsurprisingly, the German delegates named the largest number – a total of 27. The top two were the mathematicians Niels Henrik Abel and Sophus Lie, with Sophus Bugge, the linguist, coming third. There then followed the historians Peter Andreas Munch, Rudolf Keyser and Gustav Storm. It is remarkable that Ernst Sars, at that time the most well-known and influential historian in Norway itself, was not mentioned by a single foreign delegate. The reason for this is obvious, as Sars had never published anything in a foreign language.

The traditional links between the Norwegians and the German-speaking region had seen generations of Norwegian academics attending the universities of Greifswald, Rostock, Heidelberg, Leipzig, Berlin, Munich, Vienna or Göttingen to further their education. This made it natural that there was a strong German presence at the jubilee, and, in fact, German guests delivered almost 30 congratulatory speeches during the festivities. Many German speakers underlined the common cultural and ethnic origin of Germany and Norway, using terms such as "Germanic" or "North Germanic" to describe it. Conversely, terms such as "Norwegian", "Scandinavian" or "Nordic" were hardly used in any of their speeches. Kristiania and Norway were unceremoniously aligned in an imaginary common Germanic cultural sphere of influence that defined the geopolitical and cultural situation of the university in Kristiania.

This language reflected a strong renewal of interest in the North within Germany at that time, as it was believed to be the area from where the Germanic tribes had originated. Kaiser William enjoyed sailing his yacht *Hohenzollern* along the Norwegian coast, and media reports of his trips had made Norway a topic of interest among the German public.[22]

The Danish and Swedish letters of congratulation referred extensively to the national and universal characteristics of the Norwegian university and to its national identity-building functions. They made very little mention of any kind of common Nordic or Scandinavian culture, in striking contrast to the 1861 celebration of the Kristiania University's 50th anniversary, when Scandinavism had been in focus.[23]

22 See Paul Guessfeldt, *Kaiser Wilhelms II's Reisen nach Norwegen in den Jahren 1889 bis 1892* (Berlin: Gebrueder Paetel 1892).

23 John Peter Collett, "The Christiania University's 50 Years Celebration in 1861. National Pride and Scandinavian Solidarity", in: Dhondt (ed.), *National, Nordic or European?* (2011): 73–98 and Friedman, "Science, Populist Democracy and Honour", in: Dhondt (ed.), *National, Nordic or European* (2011): 267–284.

Scandinavism had been a forceful cultural and academic movement in the nineteenth century, but at the beginning of the following century, it had suffered a severe blow and had barely survived. In an interview during the 1911 jubilee, the rector of the Copenhagen University, the professor of history Kristian Erslev, deplored the fact that previous efforts to achieve a closer cooperation between Scandinavian universities had come to a standstill, the result of the unfavourable political atmosphere created by the struggle between Norway and Sweden before the dissolution of the union between them in 1905, and it had taken time for the wounds to heal.

However, amongst the more general chorus of good will, there also happened to be a discernible discord. Iceland, which had once been a part of the Norwegian monarchy but which was now under the Danish crown, was represented at the jubilee by Björn Magnusson Olsen, the first rector of the newly founded Icelandic university in Reykjavik. However, he had not been allowed to speak publicly. The various foreign delegations had been grouped according to national and linguistic criteria, defined by the Norwegian Ministry of Foreign Affairs, who insisted that this was in accordance with customary diplomatic protocol. Each group was allotted one speaker at the ceremony on 4 September. Thus, France represented the whole Latin group; Russia spoke for Finland, and so on. The fact that Iceland was not allowed to deliver its own speech, but had to be represented by the Danish delegates at a time when the Icelanders were struggling for their autonomy, was felt by many Norwegians to be an embarrassment and an insult to their Icelandic brethren. A series of readers' letters in the newspapers revealed that a deep wound had been reopened. The young Icelandic university, founded only three months before, had not been allowed to greet the Norwegian university in its own Old Norse language! This was strongly resented as a gesture of submission to the Danish oppressor.

In the newspapers, especially those published on the West coast of Norway, where nationalistic and anti-Danish sentiments had a strong foothold, this arrangement was condemned as outrageous. In a speech to the Student Union in Kristiania (*Det Norske Studentersamfund*) the nationalistic intellectual Jacob Macody Lund sharply criticized the university and rector Brøgger in particular. Lund declared that he was ashamed on behalf of the Norwegian nation for having stooped before the Danes in this way.[24] Brøgger gave a conciliatory reply through *Aftenposten*, saying that no insult had been intended

24 A report of Macody Lund's speech appeared in the national newspaper *Ørebladet* (26/11/1911).

in the decision to allow the Danes to speak on behalf of their sister nation, adding that Copenhagen was, after all, the oldest and largest of the two universities.[25]

Even if such discord had been brought to the public's attention, the leaders of the Kristiania University had reason to be pleased with the outcome of the jubilee festivities. They had seized the jubilee as an opportunity to gain recognition for everything the Norwegian university had achieved, to publicize its successes and to aim for a more prominent place in the public sphere, and they had no reason to be disappointed.

The university had not always been seen as a source of national pride. In the years between 1870 and the mid-1890s, the university had, in fact, shown a very ambivalent attitude towards the emerging liberal and democratic political movements. In the struggles over the constitution which led to the triumph of the parliamentary system in 1884 and, subsequently, to the dissolution of the union with Sweden twenty years later, protagonists for both sides were found among the university's professors. Nevertheless, the progressive forces, represented by the liberal-democrat party *Venstre*, had during those years accused the conservative professors of law, theology and philosophy of siding with the reactionaries and of having abused their authority. The accusations were not totally unjustified, and the *Venstre* majority in parliament had reacted by terminating some professorships and imposing new, radical professors on the university, among them the left-wing liberal historian Ernst Sars and the historian of folk culture Moltke Moe.

The new university statutes, which were passed by parliament in 1905, the year when the union with Sweden was dissolved, were the result of a long series of confrontations between parliament and *Venstre* on the one side, and the government, the university and the Ministry of Education on the other. The conflict was about autonomy and self-assertion and whether the university should be allowed to control the appointments of its professors or not. It was also about academic freedom in a number of other respects. The struggle ended with a new "contract" or general understanding between the university and the state. The university's academic freedom was upheld and, in fact, strengthened, but the university at the same time lost much of its former influence on the political arena. The "professor-politician" of the nineteenth century became an anachronistic figure. In the twentieth century, a professor was

25 For more details on this affair and especially the impact of it in Iceland, see Guðmundur Hálfdanarson, "University of Iceland. A Citizen of the Respublica Scientiarum or a Nursery for the Nation?", in: Dhondt (ed.), *National, Nordic or European* (2011): 303–306.

no longer automatically a public figure with the expectation and right to have a direct influence on how the country was administered and ruled, but would find other fields of activity where he could act and be recognized as an expert. If a professor wanted to take part in politics, he would have to navigate his way through a political party, such as, for instance, Fredrik Stang, who served as a government minister and a member of parliament, and as the chairman of the conservative (*Høyre*) party.

Summarizing the above, Norway's only university had, at the time of its first centenary, a close and almost symbiotic relationship with the nation, which had come into existence shortly after the university itself. The symbiosis was highlighted during the 1911 jubilee, as observed by the journalist from Manchester. Harmony was now re-established between the powers of the state and the republic of the learned, after a period of struggle and deep conflicts.[26]

The relationship between university and people, meaning the general public, appears also to have been harmonious, although a few critical voices were heard during the jubilee. Some commentators had deplored the lack of a popular spirit, others the lack of national self-consciousness. The university was not able to escape completely from the cultural struggles dividing Norwegian society, with the questions of national identity and language at their core. In the eyes of the most fervent nationalists, the university was still perceived as a symbol of the centuries-long imposition of Danish culture and language in Norway, which the forces in the nationalist *Norskdom* movement wanted to break away from.

Nor was the university above the political and social struggles of the day. The socialist press was divided into two opposing camps, however. The newspaper of the social-democrats was very critical, asserting that as long as it continued to recruit its professors and students mainly from the educated bourgeoisie and the rich, the Kristiania university could not possibly be recognized as the carrier of universal truth. Such a university would only represent the truths, opinions and values of the privileged few. The other socialist newspaper, *Demokraten,* commented on its part that as the university was open to everyone and there were no tuition fees, it was just a question of time before the gifted sons and daughters of the working class would also go to study there.

26 A concise history of the University of Oslo is to be found in John Peter Collett, *Historien om Universitetet i Oslo* (Oslo: Universitetsforlaget 1999). Another work in 9 volumes, describing the foreground to the foundation of the university in 1811 and the following 200 years was published in 2011: John Peter Collett (ed.), *Universitetet i Oslo 1811–2011*, 9 vols. (Oslo: Unipub forlag 2011).

Consequently, this newspaper was optimistic about the benefits that science and learning could bring to society as a whole.

The journey to Bergen and the various celebrations around the country had made the jubilee look like an important national event, widely supported all over Norway. The subordinated position, to which the Icelanders, Norway's former countrymen, had been relegated, caused some disagreement as we have seen. However, the protests that the university had not freed itself from Danish cultural power were not so insistent, and the contention that the university was merely a provincial outpost in a sphere dominated by German culture was not felt to be justified.

And after the Jubilee? A Retrospective View on the Position of the University

The jubilee had demonstrated that the Norwegian academics lacked neither national consciousness, nor self-confidence. Brøgger's hand could be felt behind the proceedings. His goal of making the university's role better understood and obtaining more money for research (through his historic account of the university, among other means), enabled him to point out the faults and weaknesses of the university, but also, at the same time, to outline the visions of the future to be shared by the people and its political leaders.

Brøgger's ability to imagine great, grandiose, even impossibly ambitious projects such as a world research institution, made its impact. The jubilee year was the impetus for the spectacular growth experienced by the Kristiania University in the following decade. In the form of new buildings and enlargement of existing scientific installations, in the number of professorships and other posts, and in the enrolment of students, the university in 1920 had reached double its size of 1911, and growth continued even in the economically difficult years that followed. The development in Kristiania was parallel to what happened elsewhere in the world, notably in the USA and in Germany, where there was also enormous growth, following the model of the research university.

Looking at the university in Kristiania in retrospect from Oslo in 1925, we can ask whether it had indeed developed in the direction that Morgenstierne had proposed, or more like Stang and Brøgger had wished. If we focus on student enrolment, we find that the university had expanded in three directions – socially, geographically and gender-wise. Students now came from all parts of the country and from almost all social backgrounds, and included men and women, as Morgenstierne had expected. It was far from a mass university in a modern, post-1960s sense, but the expansion was significant and tangible. Moreover, student recruitment on this wider basis actually fitted in quite well with the ideals of the research university. According to Brøgger, neither scientific talent nor research as a vocation was elitist in the sense that only students from elite backgrounds should have access to the university. On the contrary, Brøgger insisted: talent is distributed equally among all layers of the population – diligence and talent should overcome a modest social background. There were real-life examples of this: world-famous Norwegian scientists like the economist Ragnar Frisch, the astrophysicist Svein Rosseland, the mathematician Theodor Skolem, as well as a number of other scientists who made outstanding careers at the University of Kristiania/Oslo in the interwar period, were recruited to the university from outside established academic or economic elites. In this respect, the Norwegian university stood out as remarkably open, compared to other universities in Europe. A career such as Rosseland's, who had grown up on a very modest farm, would probably have been unimaginable in Germany at the same time: at the age of 21 he had been able to go to Copenhagen to study with Niels Bohr and in 1934, returning to Oslo from a professorship at Harvard and aged 40, he had become the founder and director of the first specialised astrophysics institute in the world.[27]

What does the university need most: able teachers or excellent scholars? Proponents of "the people's university" and "the research university" would tend to disagree in principle but, in reality, the way the Kristiania University actually recruited its professors and lecturers is more nuanced. The university, in fact, felt compelled to take both research and teaching seriously. In many cases, better teachers were preferred precisely for their ability to teach, but if there were excellent scholars waiting for promotion and no ordinary chair was available, in many cases personal professorships were created to keep outstanding people at the university. This was the case, for example, with the mineralogist Victor Moritz Goldschmidt and the Nobel Prize winning chemist Odd Hassel, and also with Kristine Bonnevie and Ragnar Frisch. In 1937, Professor Didrik Arup Seip, on his inauguration as rector, went as far as to say that this had become the guiding principle of recruitment to Oslo University professorships: the university had insisted that chairs be offered to researchers of outstanding qualities that it wished to recruit onto its staff, and not be constrained by the needs of education.[28]

---

27 About Niels Bohr and Svein Rosseland, see Peter Robertson, *The Early years. The Niels Bohr Institute 1921–1930* (Copenhagen: Akademisk Forlag 1979) and Østein Elgarøy and Øivind Hauge, *Svein Rosseland. Fra hans liv og virke* (Oslo: Institutt for teoretisk astrofysikk 1944).

28 Didrik Arup Seip in *Samtiden* (1937).

What was given priority: participation at an international level in specialized scientific fields, or the task of educating the people? The answer to this is also both. Some academics, notably the zoologist Bonnevie, professor of archaeology Anton Wilhelm Brøgger, and the professor of history Halvdan Koht, were able to fulfil both tasks. Others concentrated on building up their specialist networks and seldom appeared in public, and yet others were constantly on the public stage, for instance Professor Francis Bull, who was much loved for bringing an appreciation of literary history to a wider public. Towards the end of his life, long after he had retired, he even became the university's first television star.

As regards the last criterion for a research university, namely its organizational and material structure, was the university's leadership able to assure adequate space, buildings, laboratories, equipment, support staff and money for research materials, travel and publications? Indeed, a new natural sciences complex was set up in Blindern, and by 1936 this complex was housing four new ultramodern institutions: the Institute for Astrophysics – financed by the Rockefeller Foundation – and others backed by private and state funding. However, the physics and chemistry buildings were soon overrun by other departments that had been driven out of the old university building in the city centre. Consequently, what had been planned as spacious research laboratories were severely restricted by the needs of teaching. Also, in the economic depression of the 1930s far too many talented scholars left Norway due to lack of research funding and job opportunities, to find better conditions in Germany and, more especially, in the USA.

Conclusion

As we have seen, ideals and realities converged to combine the characteristics of the people's university and the research university in Kristiania/Oslo in the years following 1911. The visions brought forward at the centenary jubilee through speeches and (historical) publications came true, at least to some extent. Student recruitment expanded, and new, modern installations were built for research. Investment in research yielded scientific breakthroughs – even compared to the highest international standards – which were achieved at the university, both by individuals and teams of researchers. However, during the interwar years, the university as a whole was characterized by a lack of funding and opportunities, and much that had been gained in the decade from 1911 to 1920 was lost during the economic hardship in the years to follow. The student population doubled between 1920 and 1940, but the academic staff barely increased during the same period, and the realities of the people's university clashed with the ideals of the research university. Faculty was overburdened with teaching, conditions for research deteriorated, and recruitment of new talent came to a virtual standstill.

Returning to the question in the title of this paper: was the Royal Frederik University in Kristiania an intellectual beacon of the North, or was it a "North Germanic" provincial university? With hindsight, in view of the developments that followed, the first characteristic is clearly the more visible. The university in Kristiania, later Oslo, was able to continue its growth as an institution for higher education and training with its own specific identity, as an institution that contributed significantly to the cultural and social development of Norwegian society. In some fields it was also able to contribute to research at an international level: we have already mentioned the professors Goldschmidt, Rosseland, Hassel and Frisch.[29] We could further mention the biblical scholar Sigmund Mowinckel and a strong group of linguists, including Sten Konow and his successor Georg von Munthe af Morgenstierne, who gained international recognition for their works on Indian and Iranian languages and whose names are still cherished in Kabul today.

Would it have been just to regard the Royal Frederik University in Kristiania at the beginning of the twentieth century as a Germanic provincial university? In 1911, Germany was still the most important foreign country for Norwegian scholars and academics in general, but the Norwegian university already had a long history with its own particular characteristics: the university had developed on Norwegian soil, closely linked to the Norwegian state and the Norwegian people, identifying itself strongly with the Norwegian nation. The German universities, rising to become the undisputed world leaders in the nineteenth century, had made their impact on the Kristiania University as on most other universities in the world.[30] However, the Kristiania University – like Norwegian society and culture at large – retained many characteristics, which differentiated it from the German university and, specifically, from German elite culture.

The modernized and liberalized Norwegian university as it emerged in 1911, tightly committed to the hopes and ideals of the nation and the people, bore only little resemblance to the predominantly conservative and state-dominated universities of the German Kaiserreich. Waldemar Christopher Brøgger was inspired by the German KWG in his plans for the expansion of

29 Both of the latter were awarded Nobel Prizes, Hassel for chemistry and Frisch the Nobel Memorial Prize for economics.

30 Schwinges (ed.) *Humboldt International* (2001).

research, but both Brøgger himself and other leading university thinkers in Norway also formulated other plans for the future, in which German models played a minor role.

Academic relations with other countries were intensified up to 1911, and they were to gain further importance during the following years. In France, Norwegian historians and linguists encountered new ideas. Ultimately, from the turn of the century it was the USA that attracted Norwegian scholars most and they increasingly travelled there. Brøgger was one of the first, and although he never mastered the English language himself, he was quick to realize the historic turning point marked by the USA's rise to power. The centre of gravity of the international academic community moved away from the German-speaking area towards the emerging scientific superpower in North America. As we have already mentioned, the American academic community also stretched out an open and inviting hand to their Norwegian counterparts. The United States delegation to the 1911 jubilee in Kristiania was the largest and counted a number of academics of Norwegian descent, who had come to honour an institution of such great importance for the country of their forefathers. Also, a large part of the financing of the new university Aula was made possible by contributions from Kristiania University alumni who had immigrated to the USA.

In Kristiania, the Norwegian-American academics were pleased to find a university committed to ideals of democracy and equality that could not be taken for granted in all European countries. Conversely, when Norwegian scholars and men of letters went to the United States, they were impressed by the modernity of American society, where science and technology bore the promise of realizing the American dream of economic, cultural and social progress, and where philanthropists like John D. Rockefeller had turned billion-dollar fortunes over to the scientific community in order to assure the financing of research, independently of state control and political interference, the very conditions that Brøgger had always dreamt of.

After the First World War, Germany and its academic institutions were largely discredited in Norwegian eyes. Instead of going to Germany as in the nineteenth century, aspiring Norwegian scholars went to America. With the promise of Rockefeller Foundation grants, they came back to Norway and the University of Oslo in the 1930s, with plans to build a Norwegian university as well as a Norwegian society strongly influenced by the modern ideals of the United States.

CHAPTER 4

Commitment, Reserve and Self-Assertion

The Celebration of Patriotic Anniversaries in Russian and German Universities 1912/13[1]

Trude Maurer

Russian universities were modelled on the German reform universities of the eighteenth century and developed in a way similar to the German research university that evolved during the nineteenth century. Many future professors were (partly) trained in Germany and shared the German concepts of Wissenschaft*, academic ethics and a constitutional monarchy. Despite this common ground, the position of academics in their respective societies and their attitude towards the political system of their home countries differed considerably. To what extent they committed themselves to the patriotic cause or rather remained aloof is demonstrated by a comparative analysis of the commemoration of the anti-Napoleonic wars and the celebration of dynastic jubilees in 1913. The differences can partly be explained by their respective state's support or interdiction of university jubilees shortly before. At the same time, the findings help to explain the differences of participation of German and Russian academics in the war efforts of 1914–1918.*

In Russia, the early twentieth century was a period of cancelled, postponed and even officially banned university jubilees. Only the Russified university in the Baltics, Jur'ev (formerly Dorpat) was able to celebrate its centenary.[2] In Moscow, they had started to plan the 150th anniversary celebrations to be held in 1905 as early as 1898, but had been declined permission. The Ministry of

1 This comparative essay is based on two comprehensive articles on the Russian and German celebrations, respectively Trude Maurer, "Engagement, Distanz und Selbstbehauptung. Die Feier der patriotischen Jubiläen 1913 an den deutschen Universitäten", *Jahrbuch für Universitätsgeschichte* 14 (2011): 149–164; Trude Maurer, "Distanz und Selbstbehauptung: Die patriotischen Jubiläen des Studienjahres 1912/13 als Brennspiegel der Gesellschaftsgeschichte russischer Universitäten", in: Matthias Stadelmann and Lilia Antipow (eds.), *Schlüsseljahre. Zentrale Konstellationen der mittel- und osteuropäischen Geschichte. Festschrift für Helmut Altrichter zum 65. Geburtstag* (Stuttgart: Franz Steiner 2011): 233–254. For detailed references see these articles. Only the introductory paragraphs concerning university jubilees are completely new and have therefore been documented more fully. For other sections only quotations from or summaries of particular speeches will be cited.

2 A.R., "Brief aus Jurjew (Dorpat). Die Feier des 100jährigen Jubiläums der vormals Dörptschen, jetzt Jurjewschen Universität", *Russische Medicinische Rundschau* 3 (1902–1903): 294–296.

 | DOI 10.1163/9789004265073_005

Education claimed that there was no legal foundation for celebrating non-centenaries.[3] In 1905, amidst of the so-called Banquet Campaign of the Liberal Union of Liberation,[4] even the annual student banquets celebrated on 12 January were forbidden.[5] At the same time, the council of the University of Char'kov wanted to transfer the celebration of its centenary from January to October 1905 because of the Russo-Japenese war. Moreover, the authorities closed down the premises of the noble assembly where there had at least been a banquet planned for the day of inauguration. By doing so, however, they provoked a public meeting which called for a constituent assembly. In autumn the universities reopened only for a short time, and in October barricades of the 1905 revolution blocked University Street in Char'kov.[6] In Kazan in 1904 the 100th anniversary of the university's inauguration had also been postponed because of the war in the Far East. They intended to celebrate it ten years later on the centenary of the university's so-called "real opening", but in 1914 had to defer it again, as another war had just begun.[7] Thus, at the beginning of the twentieth century, nearly all Russian university jubilees were, for political reasons, put off *ad calendas graecas*.

In contrast, all Russian universities contributed to the celebrations of the patriotic anniversaries of 1912–1913. At the end of August 1912, the 100th anniversary of the Battle of Borodino either opened or just predated the academic year. At the end of February 1913, the tercentenary of the Romanov dynasty followed.

---

3 In Russian: 'nekruglye daty', literally meaning: 'not round dates'. In fact, during the second half of the nineteenth century 'jubilees' (*jubilei*) were celebrated after 50 or 100 years, but at the end of the century events or particular periods of service were celebrated after 25 years (Konstantin Nikolaevič Cimbaev, "Fenomen jubileemanii v rossijskoj obščestvennoj žizni konca XIX – načala XX veka [The phenomenon of jubileemania in Russian public life at the end of the nineteenth and beginning of the twentieth centuries]", *Voprosy istorii* (2005), no. 11: 100). However, the 50th anniversary of the University of Char'kov in 1855 was not celebrated because the tsar found the "lifespan" too short (Sergej Ivanovič Posochov, "Jubilei Char'kovskogo universiteta [The anniversaries of Char'kov University]", *Otečestvennaja istorija* (2004), no. 6: 142).

4 In order to circumvent the ban on public meetings the liberal intelligentsia organized banquets in the autumn of 1904 where plans for a thorough reform of the political system were discussed. In some places even demands for a constituent assembly were made.

5 "Jubilei [Anniversaries]", in: *Imperatorskij Moskovskij universitet 1755–1917. Ėnciklopedičeskij Slovar'* (Moscow: ROSSPĖN 2010): 866–869.

6 Posochov, "Jubilei Char'kovskogo universiteta" (2004): 143.

7 Elena Anatol'evna Višlenkova, Svetlana Jur'evna Malyševa and Alla Arkad'evna Sal'nikova, *Terra universitatis. Dva veka universitetskoj kul'tury v Kazani* [*Terra universitatis. Two centuries of university culture in Kazan*] (Kazan: Kazanskij gosudarstvennyj universitet 2005): 440.

In Germany, by contrast, a series of university jubilees during the first decade of the twentieth century appeared to be continued by the patriotic celebrations of 1913. Between 1907 and 1910, the universities of Gießen, Leipzig and Berlin had celebrated their own jubilees with great aplomb, including the respective monarch's presence and delegates from many universities.[8] Early in 1913, the universities (as well as other institutions and towns) commemorated the beginning of the War of Liberation, in October the monument in memory of the Battle of the Nations was inaugurated with student delegations from all over Germany attending. In between, i.e. in June, the silver jubilee of Emperor William's accession to the throne was celebrated.

The contrast between Russia and Germany on the one hand, and within Russia between the academic and the patriotic anniversaries on the other hand, requires closer examination. Comparing these two countries seems particularly appropriate as the Russian institutions, created at the beginning of the nineteenth century, had been modelled on the German reform universities, in particular Göttingen, and many professors had spent a year or two in Germany while preparing for their university career.[9] What is more, in the jubilee-mania of the early twentieth century, Russian authorities used German anniversaries, in particular the bicentenary of the kingdom of Prussia, as a model for the celebration of the Romanov jubilee.[10] In 1913, many people in both countries thought that war was imminent, and sometimes scholars even mentioned that when they met, for example, at the international congress of

8 To Berlin they had even come from three continents. In each of these towns the festivities had gone on for several days. See the official reports: Erich Schmidt, *Jahrhundertfeier der Königlichen Friedrich-Wilhelms-Universität zu Berlin, 10.-12. Oktober 1910. Bericht* [...] *erstattet von dem Prorektor* (Berlin: Schade 1911); *Zur Erinnerung an die dritte Jahrhundertfeier der Großherzoglich Hessischen Ludwigsuniversität in den Tagen vom 31. Juli bis zum 3. August 1907* (Gießen: Kindt 1907); Karl Binding, *Die Feier des Fünfhundertjährigen Bestehens der Universität Leipzig. Amtlicher Bericht im Auftrage des akademischen Senates* (Leipzig: Hirzel 1910). The University of Breslau, however, skipped its bicentenary in 1902 (Norbert Conrads, "Zur Eröffnung der Ausstellung in Görlitz (Auszüge)", in: Norbert Conrads (ed.), *Die tolerierte Universität. 300 Jahre Universität Breslau 1702 bis 2002* (Stuttgart: Franz Steiner 2004): 12).

9 On the adaptation of the model of German reform universities (of the eighteenth century) and the development of universities of both countries into research universities during the second half of the nineteenth century, see Trude Maurer, "Egalität und Weltläufigkeit. Zur Modernität rußländischer Universitäten und ihrer Professorenschaft", *Comparativ* 17 (2007), no. 5/6: 146–160.

10 Konstantin Tsimbaev, "'Jubiläumsfieber'. Kriegserfahrung in den Erinnerungsfeiern in Russland Ende des 19. bis Anfang des 20. Jahrhunderts", in: Gert Melville and Karl-Siegbert Rehberg (eds.), *Gründungsmythen. Genealogien. Memorialzeichen. Beiträge zur institutionellen Konstruktion von Kontinuität* (Köln/Weimar/Wien: Böhlau 2004): 86.

historians in London, and spoke about the next congress that was planned for 1918 in Saint Petersburg.

The organization of patriotic anniversaries differed widely. In Germany they were initiated by citizens and voluntary associations. As early as 1910, the Prussian historian Friedrich Meinecke, at that time a professor at Freiburg (in Baden), had suggested that all over Germany the universities and other institutions cooperate for the centenary of the War of Liberation. Not having succeeded, he put forward another idea in 1912: the celebrations held at individual universities should work like one big unified ceremony of German professors and students. Both ideas, however, were rejected by Prussian and Reich authorities. The varying design of local celebrations proved that there was no uniform guideline even for Prussian universities.

In Russia, by contrast, the central Ministry of Education on both occasions invited suggestions for the celebrations and then issued detailed guidelines which applied to all schools as well as institutions of higher education.

Regarding the War of Liberation, it was natural that German professors and students should commemorate the first war in which both these groups had taken part in as they had responded to the call-up of volunteers issued by the Prussian king in his famous plea "To My People". Since the centenary of the University of Berlin in 1910, this had become a central part of its image, whilst some non-Prussian universities, in turn, emphasized their own contribution.[11]

However, commemorating a national uprising in the university, which as a centre of learning was considered supranational, implied an inherent tension. Max Lehmann, a Prussian historian at Göttingen, appreciated what the Germans owed to other nations in their successful fight against Napoleon: to the Russians, the English, the Spaniards and even the French. The latter, after all, had given them the idea of compulsory military service for all citizens.[12]

Others however, though they rejected chauvinism, emphasized that the aim of their reflections was not a weakening of the patriotic attitude, but rather the preservation of its momentum. At the Berlin celebration, the pan-German historian Dietrich Schäfer even claimed: "Gaining an insight can only strengthen and purify patriotic feeling." According to Schäfer, patriotic attitude and scholarly world-view had fused completely in 1813. In the final analysis, the Berlin ceremony did not aim at reflection but at mobilization. If the Empire had to

11 The idea that the War of Liberation from Napoleonic occupation had also been a war for liberty (i.e. for a unified and democratic Germany) was put forward at various universities, but considered a legend by at least one liberal speaker in Southern Germany.

12 Max Lehmann, *Die Erhebung von 1813. Rede zur Feier des Gedächtnisses von 1813* (Göttingen: Vandenhoeck & Ruprecht 1913): 19.

win another struggle for survival, the German universities would act as the Prussians had done in 1813, Schäfer assured.[13] The student representative reaffirmed this, whilst the rector swore to educate the students on becoming competent citizens and to preserve the spirit of the forefathers: to be ready to sacrifice one's life, if necessary. Such vows were undertaken in other universities as well, but it was only the Berlin celebration which was attended by the Emperor, who even delivered a speech himself. Both there and in Breslau, the ceremony ended in a song by Ernst Moritz Arndt about the God who "had made iron grow" and did not want slaves. It culminated in the appeal "All you Germans, every single one, unite for holy war!"[14]

In all these ceremonies, the German 'people' had been the focus of attention. The Kaiser's silver jubilee emphasized the unity of the monarch and his people, and this time the universities even united in a joint tribute. Originally, this had been an idea intended only for the Prussian universities and was then expanded to include all German ones. However, it became obsolete because the University of Berlin had already decided to submit its own address and thus provoke the plan for the joint action of non-Prussian universities. Eventually, it became an all-German action with the exception of Berlin, but as soon as the Emperor had agreed upon receiving the rectors, their Berlin colleague invited them to his university, thus uniting with them whilst alos initiating the lead.

In regard to the Kaiser, some professors were truly uncritical, even ridiculously so. Others, however, were more discriminating and discussed various fields of policy. At the Berlin ceremony, the historian Otto Hintze revealed that William II had torn to pieces the political testament of Frederick William IV in which the latter had obliged his successors to violate the constitution as soon as they ascended to the throne. In this way, the Kaiser (who seems not to have been present at the time) appeared as a defender of the constitution.

In a number of addresses, the speakers cautiously distanced themselves from the emperor, not in principle of course, but regarding particular aspects. Most clearly the distance was marked by Friedrich Meinecke: "Our emperor's fate is our fate [...]. Our perceptions and judgements may often have differed, because we serve the monarchy being free persons. But we will not leave him

---

13 Schäfer's speech is documented in: *Feier der Königlichen Friedrich-Wilhelms-Universität zu Berlin* [...] *zur Erinnerung an die Erhebung der deutschen Nation im Jahre 1813* (Berlin: Schade 1913): 5–29.

14 "Der Gott, der Eisen wachsen ließ, der wollte keine Knechte [....] Ihr Deutschen alle, Mann für Mann, zum heil'gen Krieg zusammen!" (quoted in: *Feier der Königlichen Friedrich-Wilhelms-Universität zu Berlin* (1913): 34).

nor will he leave us. Together we serve that great ideal of nation [...]." It was on this basis that Meinecke also assured his willingness to join forces in the future. "The real battlefields of our time are still before us," he said. And therefore this was not the time for boasting or for making a great show. Meinecke reiterated the "serious decision to live and die for", though in this case the statement referred to the "genius of our people".[15] Obviously, there was a loyalty superior to that which they owed the monarch.

The willingness to go into battle was guaranteed at a number of celebrations and, to those participants, this did not contradict in any way William's image as an emperor of peace. Rather both elements fused. Otto Hintze justified the continued armament as an "insurance premium" for the preservation of peace. But to Hintze this peace seemed to hold little attraction, as he continued: "And though we have not been spared the experience that nothing is harder to endure than a series of good days, this time has been filled by unremitting honest working." In a long period of down-to-earth hard work, the Germans had been able to maintain and develop what had been created 43 years earlier "in the storm and stress of the enthusiasm of war".[16] Students who took vows at these celebrations or at their own meetings invoked their long tradition for battle.

No such declarations of willingness can be found in the records of Russian universities. The commemoration of the Battle of Borodino surpassed all former anniversaries (not only military ones) by the variety of celebrations and the way in which they spread all over the empire. In this respect, it also anticipated the tercentenary of the Romanov dynasty. The purpose was clear: to overcome the Russian defeat in the war against Japan mentally and to reconcile the people with the state machinery and the dynasty by gathering around symbols and ideas suggested by the government. Intellectuals, of course, responded reluctantly or not at all.[17]

The elements laid out in the official guidelines of the Ministry of Education included religious services, a ceremony in the assembly hall, decoration and

---

15 Friedrich Meinecke, *Festrede zur Jahrhundertfeier der deutschen Erhebung und zur Kaiserfeier* [...] *in der Aula der Universität Freiburg i. Br.* (S.l. s.d.): 13, 16.

16 Hintze refers, of course, to the empire embodying the unification of Germany. His speech is documented in: *Feier des Fünfundzwanzigjährigen Regierungs-Jubiläums Seiner Majestät des Kaisers und Königs* (Berlin: Norddeutsche Buchdruckerei 1913): 5–38.

17 On the celebration of this anniversary see Kurt Schneider, "100 Jahre nach Napoleon. Rußlands gefeierte Kriegserfahrung", *Jahrbücher für Geschichte Osteuropas* 49 (2001): 45–66; Richard Wortman, *Scenarios of Power. Myth and Ceremony in Russian Monarchy. Vol. II: From Alexander II to the Abdication of Nicholas II* (Princeton: University Press 2000): 431–438 (both without reference to the universities).

illumination of buildings, the renaming of premises in commemoration of heroes and events of the so-called Patriotic War. In addition, the institutions were asked to buy books on the subject, confer medals and set up new scholarships. In places where military parades were staged, the participation of students was "desired". (The Russian term '*učaščiesja*', as the English term 'student' does, covers both pupils and students at institutions of higher education.)

Whereas religious services and the renaming and setting up of stipends took place everywhere, the academic ceremonies varied. The University of Jur'ev, for example, gave up its plan for lectures on the events of the war, the activities of the university, and the state of medicine at that time, and organized a joint celebration with other educational institutions instead, as "permitted" by the Ministry in order to give it more solemnity. The lectures were planned to be published in the university's scholarly journal, but in fact were never released.[18]

Kazan University chose a historian who was a member of the Constitutional Democratic Party as keynoter, one who was critical of the establishment. His lectures had even been under surveillance by the secret police. In 1912, discussing the so-called people's and guerrilla war had been permitted for the first time. But this historian shattered the perception of the Patriotic War, officially conceived as a war of liberation from French oppression. He depicted it as a popular uprising that was also directed against their own leaders, drawing attention to the people's brutality and bestiality.[19]

The Moscow ceremony was organized together with the Historical Society and included four papers on various topics, including a portrayal of Alexander and Napoleon, the experience of Moscow during that war, and the depiction of 1812 in Russian 'belles lettres'. Here, a young critical lecturer also brought out the violence and rapacity of the mob. But by far the longest (and, to quote the liberal press, "long-winded") lecture was given by the rector, a conservative historian, who discussed Moscow University at the time of war.[20]

---

18 For the university's Lutheran church a plate was planned in commemoration of the fallen members of the university.

19 Nikolaj Nikolaevič Firsov, *1812 god v sociologo-psichologičeskom osveščenii. (Obščaja charakteristika)* [*A sociological and psychological interpretation of 1812. A general description*] (Moscow: Tipografija G. Lissnera i D. Sobko 1913). This was an expanded version of Firsov's lecture delivered on 26 August 1912.

20 The four speeches by Vladimir Ivanovič Ger'e [Guerrier], Sergej Vladimirovič Bachrušin, Matvej Kuz'mič Ljubavskij and Konstantin Vasil'evič Pokrovskij were published in: *Čtenija v Obščestve Istorii i Drevnostej Rossijskich* IV (1912): 5–130. A short summary may be found in: Maurer, "Distanz und Selbstbehauptung" (2011): 242. The quotation is taken from *Russkie Vedomosti* (23/10/1912), no. 244: 4.

The celebration of the tercentenary of Romanov rule aimed at uniting the sovereign with the people.[21] The guidelines suggested talks and excursions as a preparation, religious services and academic ceremonies on the anniversary, and a special design for commendations and decorations conferred in 1913. Again, the most loyal and state-oriented was the ceremony in Moscow, which comprised of four lectures and was attended by high officials as well as other honorary guests.

But on this occasion, what did not take place was even more telling than the ceremonies held. Char'kov University had no ceremony of its own, but just two lectures of the Historical Society some days after the actual anniversary, while a conservative professor spoke at a number of meetings of monarchist organizations. In Kazan, the university confined itself to a church service, decoration of the buildings and the collection of money for a new icon for the university's church as the programme originally submitted had not been confirmed.

In Saint Petersburg, the university planned a joint celebration together with the Academy of Sciences but met with (unspecified) obstacles. Therefore it initiated a joint session of the councils of all institutions of higher education. They hoped that a member of the imperial family (in his capacity as an honorary member of the university) would preside over two planned talks. The famous historian Sergej Platonov was to speak about the election of the first Romanov tsar by the Assembly of the Land in 1613, and a lecture on the development of higher education during the last three centuries. Platonov was in fact able to deliver his speech despite a year before writing that the unanimous election was the source of the Tsar's power and, thus, contradicted the official version.[22] The second lecture, however, was given by a former deputy minister and member of the State Council who also presided over the celebration. There was no member of the imperial family and worse still was that the Tsar did not receive the delegation of the University that was due to hand over its congratulations (though he did receive delegations from other universities). I can only guess that he objected to the choice of professors the university had intended to send. In addition to the rector (who was a Lutheran, constitutionalist and had sympathized with student boycotts) it also included a former

---

21 On the celebration of the tercentenary (not including the universities) see Wortman, *Scenarios of Power* (2000), vol. 2: 439–480.

22 Sergej Fedorovič Platonov, "Vsja zemlja [The whole land]", in: Platon Grigor'evič Vasenko, Sergej Fedorovič Platonov and Eena Filimonovna Turaeva-Cereteli, *Načalo dinastii Romanovych. Istoričeskie očerki* [*The origins of the Romanov dynasty. Historical studies*] (Saint Petersburg: Ja. Bašmakov i K° 1912): 234. On the official version and different interpretations see Wortman, *Scenarios of Power* (2000), vol. 2: 442.

member of the state council and former rector who had resigned from both offices in protest against political and police measures.

The most remarkable celebration was held at Jur'ev. It had been prepared by a commission which included a Lutheran professor of theology, and comprised both Lutheran and orthodox services. The delegation which was sent to Saint Petersburg included not only a Russian left liberal, but also a Baltic German professor of medicine. Moreover, Jur'ev was the only university of the whole empire where student corporations had existed for nearly a century. In the morning, corporations of various national groups, as well as pupils, soldiers and inhabitants, gathered on the town square. Thus, the celebration in Jur'ev was not a purely Russian affair and the ceremony held in the university's assembly hall in the evening was attended by the mayor, deputies of the town council and various societies. For the professor of Russian history, Ivan Lappo, the election of the first Romanov tsar did not only mean the end of the Time of Troubles, but also the formation of the new Russian state. Therefore the tsar had to be elected by the whole country, and this assembly had transformed itself into a "real representative body of the people".[23] His legitimacy was based on kinship, the grace of God and election by the Assembly of the Land. Thus, Lappo implicitly contradicted the official interpretation and designed a quasi-constitutional foundation (and limitation) for the ruling dynasty.

In the official reports on the celebrations, the universities were hardly mentioned, nor did several of them even mention their ceremonies in their own annual reports. Not having been able to celebrate their own jubilees a decade earlier, they were none too eager to do so for the imperial celebrations. Being state institutions, however, they could not boycott them. Therefore, they either confined themselves to a minimum or used the occasion for critical reappraisals. Even in the most loyal celebration a spectrum of interpretations was presented.

In German universities which, on the whole, were far more state-oriented than their Russian counterparts, there was also more thoughtfulness than expected. This similarity was the consequence of their profession and the scholarly ethos which German and Russian professors shared. There were other, rather ideological parallels. The preference for a constitutional monarchy was

23 "predstavitel'stvo naroda dejstvitel'noe"; Ivan Ivanovič Lappo, "Reč' proiznesennaja 21 fevralja 1913 goda na Toržestvennom Akte Imperatorskogo Jur'evskogo Universiteta posvjaščennom čestvovaniju 300-letnego jubileja Carstvujuščego Doma Romanovych [Talk delivered on 21 February 1913 at the Solemn Celebration of the tercentenary of the Reigning House of the Romanovs at Jur'ev University]", *Učenye Zapiski Imperatorskogo Jur'evskogo universiteta* (1913), no. 2: 18.

obvious. And in both countries the universities displayed more reserve at the monarchic anniversary than in the commemoration of the anti-Napoleonic war. Moreover, both Russian and German universities used the celebrations to present their own institutions.

The most important difference was the scope of celebrations. In Germany they were on the agenda of the whole university, whereas in Russia students hardly participated. This is more telling, as students had used many academic ceremonies as occasions for revolutionary festivals of their own since the late nineteenth century.[24] And this in turn raises the question of whether professors perhaps tried to forestall such transformations by selecting particular dates or premises for the celebrations or by associating the university with other institutions in a joint ceremony.[25]

While both students and professors in Germany committed themselves to serving in future wars, this was completely beyond the imagination of the Russian academic community. In Germany during the nineteenth century, military service had become a key element of student identity, and as it was compulsory and universal they shared this experience with their teachers (of whom most were, in fact, reserve officers). In Russia, on the other hand, military service was postponed until students completed their education and were entitled to a much shorter period of service. Teachers, including those only entering upon an academic career, were completely exempt. Whereas in Germany the military and the academic worlds overlapped, in Russia they were completely separated. While some of the vows taken may have been rather hypothetical, Hintze's remarks suggest that there was also a certain longing to prove oneself.

When the First World War broke out a year later, the two academic communities were alike only in their patriotic declarations. In Russia, teaching went on as usual. Students were only drafted in 1916, and with the exception of Jur'ev, professors did not even go to the front as medical consultants. A wider commitment on the so-called home front began only with a year's delay and was in

24 For such transformations of the university's annual ceremony and a number of funerals of Saint Petersburg professors see Evgenij Anatol'evič Rostovcev, "Revoljucionnye kommemoracii v Peterburgskom universitete na rubeže XIX–XX vv. [Revolutionary commemorations in Saint Petersburg University at the turn of the century]", *Klio* 4 (2011), no. 54: 89–99.

25 Courses could begin anytime between 20 August and the end of September, but it has proved impossible to establish the exact date for 1912 (when Firsov's reappraisal was read out at the Kazan ceremony on 26 August). For her efforts in checking, I am indebted to Jana Rudneva (Kazan). In 1913, the organizers in Saint Petersburg were obliged to hold their common celebration not in an institution of higher education.

fact not really voluntary but rather called for by state authorities. In Germany, both students and professors were liable for military service, and many even volunteered (including those who were beyond draft age). Some professors even appealed to the students to put their vows into effect there and then. And the famous theologian Adolf (von) Harnack found in Arndt's song the "superior justification for war".[26] The interpretation of World War I as a defensive war which was forced on Germany had been foreshadowed by the combination of love for peace and preparedness for battle in the 1913 ceremonies.[27] The celebration of the anniversary of the War of Liberation had anticipated the military commitment both mentally and emotionally and, in turn, strengthened it in 1914.

26 From a student's lecture notes quoted in: Christian Nottmeier, *Adolf von Harnack und die deutsche Politik 1890–1930. Eine biographische Studie zum Verhältnis von Protestantismus, Wissenschaft und Politik* (Tübingen: Mohr Siebeck 2004): 378. The classical philologist Eduard Schwartz also recalled Arndt's song and drew a line from the War of Liberation through the war of 1870 against France to World War I in his speech on 24 October 1914 (the first in a Strasbourg lecture series on the war): E[duard] Schwartz, *Der Krieg als nationales Erlebnis* (Straßburg: Trübner 1914): 4, 9. For a specific lecture on this tradition of national commitment and its culmination in World War I see the next reference.

27 H[arry] Breßlau, *1813. 1870. 1914. Rede gehalten im Saal der Aubette zu Straßburg am 31. Oktober 1914* (Straßburg: Trübner 1914): 2. Breßlau was a medievalist of Prusso-Jewish origin. Also see Meinecke's article in which he brings out both the "individual character" of the "national uprisings" of 1813, 1848, 1870 and 1914 and claims that they were part of the "continuity of national life": Friedrich Meinecke, "Die deutschen Erhebungen von 1813, 1848, 1870 und 1914", in: Friedrich Meinecke, *Die deutsche Erhebung von 1914* (Stuttgart: Cotta 1914): 10.

CHAPTER 5

Academic Ceremonies and Celebrations at the Romanian University of Cluj 1919–2009

Ana-Maria Stan

This case study is dedicated to an academic institution set in a multicultural environment in Eastern Europe, in the city of Cluj/Kolozsvár in Transylvania. Founded in 1872, during the Austro-Hungarian period, from 1919 onwards Cluj University has functioned under the administration of the Romanian state. The present article investigates the diverse forms of festivities that shaped a specific institutional identity for this university, outlining continuities and fractures, often determined by political regimes. During the first half of the twentieth century, academic ceremonies resulted in the University of Cluj becoming a reference amongst other Romanian provincial universities, giving it an air of prestige, with a defined role on the local, regional and national public scene. Its professors and students were perceived as an elite group, with both civic and scientific responsibilities. During the communist period this situation modified dramatically. The university's main task was now to contribute, even through the festive moments, to the creation of the "new man", so highly praised in the speeches of Romanian communist leaders. Obviously, such deep and meaningful shifts of perspective were accurately reflected in the way Cluj university history has been written over the years by specialists or direct participants. The search for an academic individuality and the setting up of long-lasting anniversary traditions were, are and will remain an ongoing process in Cluj, as each academic jubilee provided a different context of reflection on the subject, giving birth to many types of publications or in-house events. Furthermore, the ups and downs in the life of this educational institution are closely linked to the history of the city and could offer to the general and professional public an interesting image of Cluj and its elite. Consequently, our study aims to show that the Romanian University of Cluj could stand as a specific example within the larger European typology of academic jubilees and celebrations.

In his well-known and much-discussed anthology, *The Idea of a University*, first published in 1859, John Henry Newman offered an interesting vision on the role and place of universities in the modern world. Amongst his reflections, he discussed how an academic identity comes into being. Acknowledging that the main task of the university was to "educate the intellect to reason well in all matters, to reach out towards truth, and to grasp it", Newman also underlined

 | DOI 10.1163/9789004265073_006

the fact that such an institution, which brought together a "youthful community", would indisputably give birth to a "living teaching, which in course of time will take the *shape of a self-perpetuating tradition, or a genius loci*, as it is sometimes called; which haunts the home where it has been born, and *which imbues and forms, more or less, and one by one, every individual who is successively brought under its shadow*".[1] Due to the fact that university celebrations generally concentrate on bringing forward the elements of continuity and value in academic life, they represent an efficient method of creating the institution's specificity – that *genius loci/tradition* which Newman mentions. They play a considerable role in determining and/or legitimizing a certain scholarly atmosphere as well as in generating feelings of pride and belonging shared by all the alumni who attended a particular university.

Such kinds of festive events have been part of the universities' existence since their very foundation, but from the nineteenth century onwards, and particularly since its second half, European universities have organized celebratory occasions more frequently. Consequently, academic festivities and jubilees offer a generous area of investigation for researchers and, in recent years, several projects (dedicated mostly to the German and the Scandinavian regions) analysed such jubilees from a comparative and transnational perspective. Specialists were able not only to reflect upon the multifaceted interaction between the universities and their host cities, countries or regions on such occasions, but also to define standard components and lasting patterns in such anniversaries.[2]

In contrast, a systematic and synthetic analysis of academic celebrations is still underdeveloped in Central and Eastern Europe, a multicultural patchwork region, where boundaries and political regimes have significantly fluctuated throughout the late nineteenth and twentieth centuries. Here geopolitical factors, alongside social and economic ones, have deeply shaped the creation, behaviour and presence of the academic community, both within and outside

1 John Henry Newman, "Discourse six. Knowledge view in relation to learning", in: Id., *The idea of a university. Defined and illustrated* (London/New York: Longmans, Green and co 1907): 125–126 and 147, www.newmanreader.org/works/idea/discourse6.html (date accessed 20/06/2014). Italics added by the author, except for "genius loci".

2 Pieter Dhondt (ed.), *National, Nordic or European? Nineteenth century university jubilees and Nordic cooperation* (History of Science and Medicine Library 25. Scientific and Learned Cultures and Their Institutions 4) (Leiden: Brill 2011): 2–5; Thomas P. Bekker, "Jubiläen als orte universitärer Selbstdarstellung. Entwicklungslinien des Universitätsjubiläums von der Reformationszeit bis zur Weimarer Republik", in: Rainer C. Schwinges (ed.), *Universität im öffentlichen Raum* (Veröffentlichungen der Gesellschaft für Universitäts- und Wissenschaftsgeschichte 10) (Basel: Schwabe Verlag 2008): 92–93.

the walls of the university. Ethnicity and the long term process which led to the birth of national states in this area also played a key role in defining the character of the local universities and implicitly that of the provincial, national or transnational elites formed by them.

Starting from these methodological observations, our research will present and analyse – as a case study – the way in which the Romanian University of Cluj has celebrated almost a century of purpose and performances, through different types of academic ceremonies, subsequently creating and imposing its institutional presence upon the public sphere.

Since its re-foundation in 1919, the Romanian University of Cluj in the middle of Transylvania has faced several challenges. The most important of them was related to the specific circumstances of its appearance – the new academic institution took over the infrastructure from its Hungarian predecessor (the Franz Joseph University established in 1872 in Kolozsvár/Cluj[3]) and to a large extent also its scientific and didactic organization (number and type of faculties, chairs and institutes, and a small part of the staff). This heritage has not only raised the issue of adapting, developing and improving the former standards of teaching and research, but has also demanded a deeper consideration regarding the trademark of the new institution, namely of how to organize and position the Romanian university of Cluj in the academic community, both within a national and European context.

One effective way to build the identity of the new institution was through different celebrations and jubilees, each ceremony defining or emphasizing one of its trademark components. Hence, our study will investigate the different phases of the festivities that shaped a specific identity for Cluj University, outlining continuities and fractures, often imposed or determined by political regimes. The differences between the academic celebrations that took place in Cluj while Romania was organized as a kingdom, in comparison with similar events during the communist period are significant and reveal the inherent links between politics, the evolution of educational legislation, and the critical issue of university autonomy. Each anniversary provided opportunities to

3 The Austro-Hungarian Emperor Franz Joseph officially bestowed his name upon the university in Kolozsvár/Cluj only in 1881, although the institution was set up by the Hungarian parliament in September 1872, sanctioned by Franz Joseph approximately a month later. Courses started on 11 November 1872. For details about the foundation and the first years of the Hungarian university in Cluj, see Gaal György, *Egyetem a Farkas utcában. A kolozsvári Ferenc József Tudományegyetemelőzményei, korszakai és vonzatai* [*University on Wolf Street. The Franz Joseph University of Cluj: its history, ages and implication*] (Cluj: Erdélyi Magyar Műszaki Tudományos Társaság 2001): 38–47.

assess the institutional evolution, leading to a considerable array of books and articles reflecting the academic activities of professors and students. These publications also evinced the way in which in-house traditions were being created and used for the benefit of the academic community, as well as of the inhabitants of the city, the region and even the country. In many cases, political considerations or propaganda twisted the university celebrations for their own benefit, using them as perfect occasions to present messages from the authorities; offering in that way a telling example of how culture incorporates other types of values and conveys the society at large.

The Inauguration Ceremony of the Romanian University in Cluj in 1920

The setting up of the new Romanian university in Cluj was a lengthy and complex process, of which the geo-strategic and socio-political connotations surpassed the academic ones. After the dissolution of several multinational empires, which had dominated the central and eastern part of the European continent for centuries (Austria-Hungary, tsarist Russia and the Ottoman Empire), the whole region was undergoing a profound restructuring process.

On 11 November 1918 the First World War had ended on the Western front and two days later in Belgrade, Hungary signed an armistice with the Triple Entente, acknowledging temporary demarcation lines for the territories that were to remain under Hungarian control until definitive borders could be agreed upon.[4] Transylvania was one of the most disputed provinces and consequently Hungarian, Romanian and French troops remained present on its territory and performed complicated military manoeuvres in the years 1918 and 1919.[5] On 1 December 1918 more than 100.000 Romanian citizens and 1228

4 The text of the 1918 Belgrade Armistice can be found in *The American Journal of International Law* 13 (1919), no. 4: 399–402. For an in-depth interpretation of the events, see Bogdan Krizman, "The Belgrade Armistice of 13 November 1918", *Slavonic and East European Review* 48 (1970): 67–87.

5 Gheorghe Iancu, "Officiers supérieurs français en mission dans la Transylvanie des années 1918–1919", in: George Cipăianu and Vasile Vesa (eds.), *La présence française en Roumanie pendant la Grande Guerre 1914–1918* (Cluj: PUC 1997): 153–163; Fréderic Guelton, "L'armée française et la nouvelle architecture géopolitique et militaire de l'Europe 1919–1921", in: George Cipăianu and Vasile Vesa (eds.), *La fin de la première guerre mondiale et la nouvelle architecture géopolitique européenne* (Cluj: PUC 2000): 309–320; Jean Nouzille, "La Hongrie et la défense de son intégrité territoriale en 1918–1919", in: Cipăianu and Vesa (eds.), *La fin de la première guerre mondiale* (2000): 251–300.

appointed delegates of the Romanian community assembled in the city of Alba-Iulia, proclaiming the union of Transylvania, Banat and Partium with the Romanian state (the so-called "Old Kingdom"), an act that was officially accepted by Ferdinand I, the Romanian king, by royal decree on 26 December 1918.[6] Fundamental administrative and political reforms closely followed these events. Gradually Romanians managed to take over the control of the Transylvanian villages and cities.[7] Meanwhile, the Paris Peace Conference intensively debated Transylvania's final status and pondered over the most appropriate itinerary of the border between Hungary and Romania.[8] In that way Transylvania was the last province inhabited by a Romanian population to unite with Romania, after Bessarabia (former province of the Russian empire) and Bukovina (previously part of the Austro-Hungarian Empire).[9]

In this intricate context, it is no surprise that almost a whole year passed between the moment when the Romanian authorities assumed control of the existing academic patrimony of the university in Cluj (May 1919), and the inaugural festivities (30 January – 2 February 1920). The take-over was planned carefully and accomplished in several stages.[10] One of the most important

---

6 For an analysis of Transylvania's unification process, although rather coloured by a communist interpretation of the events, see Ştefan Pascu, *The making of the Romanian unitary national state: 1918* (Bucharest: Academiei Republicii Socialiste România 1988): 192–228 and 238–260.

7 The Romanian army entered the city of Cluj on 24 December 1918 and a week later the French general Henri Berthelot was received by Romanian troops. The first Romanian prefect was appointed on 18 January 1919 and the first Romanian mayor of Cluj was sworn in on 25 February 1919.

8 Jean Baptiste Duroselle, *Histoire diplomatique de 1919 à nos jours* (Paris: Dalloz 1993): 24–31; Viorica Moisuc (ed.), *România la Conferinţa de Pace de la Paris 1918–1920. Triumful principiului naţionalităţilor* [*Romania at the Paris Peace Conference 1918–1920. The triumph of the nationalities principle*] (Cluj: Dacia 1983): 254–402.

9 Bukovina's and especially Bessarabia's union with Romania were very delicate, even forbidden topics in Romanian communist historiography. Only after 1989 this could be researched thoroughly and objectively; Viorica Moisuc, *Basarabia, Bucovina, Transilvania: unirea 1918* [*Bessarabia, Bukovina and Transylvania: the Union of 1918*] (Bucharest: Dip 1996); Ioan Scurtu, (ed.), *Istoria românilor. Vol. 8: România întregită* (*1918–1940*) [*The history of Romanians. Vol. 8: The Great Romania 1918–1940*] (Bucharest: Enciclopedică 2003).

10 For a thorough description of the transformation of the Hungarian University of Cluj into a Romanian institution, see Onisifor Ghibu, *Universitatea Daciei Superioare* [*The University of Superior Dacia*] (Bucharest: Atelierele Grafice Cultura Naţională 1929): 27–66; Irina Livezeanu, *Cultural politics in Greater Romania. Regionalism, nation building and ethnic struggle 1918–1930* (Ithaca-New York: Cornell University Press 1995): 218–227; Vasile Puşcaş (ed.), *University and society. A history of higher Education in Cluj in the*

events took place on 12 September 1919, when King Ferdinand I of Romania stated in a royal decree that "from 1 October 1919 onwards the Hungarian university of Cluj would be transformed into a Romanian language university."[11] Thus, in his position of head of state, Ferdinand I became the founder of the Romanian university of Cluj. Shortly afterwards, on 3 November 1919, teaching effectively started, with a meaningful opening speech, titled "The duty of our life", by the famous Romanian archaeologist Vasile Pârvan,[12] for a public of more than 1800 newly enrolled students and the local officials. The time that elapsed leading up to the opening ceremonies of the university noticeably signals the various stakes incorporated by this event.

First, there is the strong (symbolical) link between the city of Cluj and the university. In the eighteenth and the first half of the nineteenth century, Cluj functioned as the administrative capital of Transylvania. It lost this status after 1848, but became instead its cultural capital after the foundation of the Hungarian university in 1872. Therefore, celebrating the new Romanian university meant, in a way, marking the conversion of Cluj into the Romanian cultural capital of Transylvania, and this at a delicate moment, when the Paris Peace Conference had not yet settled the definitive ownership of this much fought-over and coveted province.[13] The political implications become even more evident when taken into account that one of the proposed options was to organise the academic festivities on 1 December 1919, exactly one year after the proclamation of Transylvania's union with the Romanian state.

Second, the inauguration ceremonies were an occasion to demonstrate the worth and the qualities of the recently formed Romanian academic community, as well as of the new public administration. Already during the period as

---

20th century (Cluj: PUC 1999): 61–129; Vasile Puşcaş, *Universitate, societate, modernizare. Organizarea şi activitatea ştiinţifică a Universităţii din Cluj. 1919–1940* [*University, society, modernization. The organisation and the scientific activity of Cluj University 1919–1940*] (Cluj: Eikon 2003): 129–245.

11 This decree was published in *Monitorul Oficial* (23/09/1919), no. 126: 7202. All the source translations have been made by the author. On the same day, a similar decree was issued in order to regulate the situation of the German language university of Cernăuţi/Czernowitz, in Bukovina. See *Monitorul Oficial* (23/09/1919), no. 126: 7202.

12 Vasile Pârvan, archaeologist, epigraphist and philosopher of history, was professor of history at the University of Bucharest and member of the Romanian Academy. He specialized in the study of ancient people living in Eastern Europe – especially Thraco-Dacians and Geto-Dacians.

13 The peace treaty that finally fixed the border between Hungary and Romania and attributed most of Transylvania to the latter was signed only on 4 June 1920 at the Grand Trianon Palace in Versailles.

a Hungarian university, Cluj had hosted some impressive ceremonies: in November 1872 at the official opening, and again in 1887 and 1895, when Franz Joseph visited the institution. Obviously, the Romanian celebrations needed to differentiate themselves from and even surpass these preceding events.

The 1920 inauguration of Cluj University also offered the opportunity for a joyful tribute to the Allies who had won the First World War, and to their socio-political and ethical values. In many respects, and particularly in the academic field, Transylvania's situation corresponded to that of Alsace-Lorraine, and the Cluj festivities have much similarity to the inauguration of the French university in Strasbourg that took place on 22 November 1919, celebrating the settlement of the historical dispute between France and Germany. It was clear that the Strasbourg celebrations were organized by the French people not only as a big cultural event, but also as a significant political and national one, and in that way it certainly functioned as a source of inspiration for its Romanian counterpart.[14] The opening and inauguration of the Romanian Cluj University effectively marked the replacement of German-inspired organizational and institutional models in Transylvania with French examples – the winners symbolically substituting the vanquished.

Last but not least, this event represented the start of what had to become a new tradition. The celebrations were intended not only as a first and significant outline of how the future brand of this institution would be constructed, but also as a benchmark in a long string of similar anniversaries and jubilees.

Obviously, there were many delicate issues regarding the inauguration ceremonies that had to be settled, leading to much discussion. Professor Sextil Puşcariu, the first rector of the Romanian University of Cluj, recalls the circumstances in his memoirs:

> In order to prepare the festivities, we worked first with a committee of professors, joined by some Cluj authorities. A lot of meetings, fewer achievements. Then I sent a telegram to Bucharest, to Doctor Nicolae Minovici,[15] a man with admirable organization skills, inviting him to come as general commissioner of the festivities. He arrived, took the academic arrangements into his hands, slaved away at it, and all went perfectly. He provided the necessary glamour to the festivities, giving them a real European appearance and brought them out of the provincialism in

14 For a comparison between the university inaugurations in Strasbourg and Cluj, see Puşcaş, *Universitate, societate, modernizare* (2003): 221–232.

15 Nicolae Minovici was a physician and university professor in Bucharest, specialized in forensic medicine.

> which they were threatened to become bogged down in. With a unitary management, things went smoothly. Secondly, I took care that the other side of the festivities, the intellectual one, would be successful and thus, by complementing and helping each other, taking everything upon our shoulders, we carried out a good thing [...]. When the guests turned up, every one of them received an envelope at the train station which contained all the invitations to the formal meals and ceremonies, indicating their place in the rooms, their accommodation, which restaurant for lunch, the full programme of the festivities, etc. Consequently, we avoided the problem of people wandering around, as usually happens in our celebrations, without knowing where they should sit or eat, the time and places they have to appear at, what clothes they need to wear, etc. Not a single glitch, everything went smoothly.[16]

The highlight of the festivities was planned for 1 February 1920. The programme of that day included the classical and customary ingredients of such a celebration, with a few specific twists and accents. The day started with a solemn ceremony in the Aula Magna of the university assembling the indigenous academic staff, delegates from other national and international universities, highly placed Romanian politicians and all the foreign ambassadors of the Allied Powers and neighbouring East Central European states. The Romanian Royal Family (King Ferdinand, Queen Marie, Crown Prince Carol and his sister, Princess Elisabeth) was also in attendance, acknowledging the significance of the event and indirectly guaranteeing the support of the authorities for this new institution. After the professional oath, pledged by all the professors of the university,[17] it was time for speeches. In his much-applauded discourse, Ferdinand I expressed an influential view on the duty of this new university. Addressing the origins of the institutional concept – *Omnium scientiarum universitas* – the monarch emphasized that: "The quality, not the quantity of graduates is the important factor – to form generations of people with character, animated by large views, permeated with love of the Homeland and devotion to education and work, eager to put their energy and knowledge capital in the service of this country."[18]

---

16 Sextil Puşcariu, *Memorii* [*Memoirs*] (Bucharest: Minerva 1978): 470–471.

17 During the first half of the twentieth century, the academic staff in Romania, especially full and assistant professors, were appointed by royal decree and assimilated to highly placed public servants. Therefore they had to make an oath of allegiance to the King and the Constitution, when taking up their office.

18 *Serbările pentru inaugurarea Universităţii din Cluj, 31 ianuarie-2 februarie 1920* [*Festivities for the inauguration of the University of Cluj, 31 January–2 February 1920*] (Bucharest: Cartea Românească 1920): 16–17.

Assuming the patronage of Cluj University, and deliberately using the word "godfather" in front of the select audience, King Ferdinand decided to bestow the generous sum of 400.000 lei to the university, in order to create an institute for the study of Romanian history. It was a choice that accurately reflected the geopolitical changes in Central and Eastern Europe after the war and at the same time it spoke volumes about the mission of this academic institution. Romanians from Transylvania, no longer a minority community in a multinational empire, received the necessary means to investigate and support their culture and consolidate their country.[19] At the same time, the King's gesture clearly indicated that the promotion of national values in research and teaching had been defined as one of the core values of the new university, and the following years would reveal the extent to which this was put into practice.

Quite a similar paradigm was sketched by many of the other speakers, taking the floor after the sovereign – especially by French Italian and Spanish diplomatic emissaries and university delegates, who called attention in their short addresses to the fact that Cluj was now (re)joining the big family of Latin culture and civilization and had to bring its own specific contribution to the universal scientific and artistic patrimony. Other foreign delegates chose to focus on the religious and, implicitly, ethnic complexity of Transylvania, like for instance the Dutch ambassador, who mentioned the long-standing contacts between the protestant students of this province and the universities of his own country – that of Utrecht in particular – and expressed the wish that this kind of academic collaboration would continue under the new conditions.[20]

An interest to preserve the regional specificity while at the same time accentuating the Romanian character of Cluj University was also visible in the inaugural discourse of rector Puşcariu. Aware of the fact that most of the enrolled students were natives of Transylvania, he advocated the necessity of studying the characteristics of this province by creating special chairs, institutes and departments. Puşcariu believed that such an approach would distinguish Cluj

19 The idea of "autonomous development" for the peoples of the Austro-Hungarian Empire was first expressed by the American president Woodrow Wilson on 8 January 1918, in his famous *Fourteen Points Speech*, which stood as a basis for the Central Powers surrender at the end of the First World War. Although the final peace treaties to a lesser extent were marked by the original Wilsonian concepts, its philosophy greatly influenced the evolution of Eastern and Central Europe during the interwar years. About the long process of cultural unification of Transylvania with Romania, see Olimpiu Boitoş, *Progresul cultural al Transilvaniei după Unire* [*The cultural progress of Transylvania after the Union*] (Sibiu: Cartea Românească din Cluj 1942): 28–51 and *Clujul. Viaţa culturală românească* [*Cluj. The Romanian cultural life*] (Cluj: Ligii Culturale, secţiunea Cluj 1929): 60–100.

20 *Serbările pentru inaugurarea Universităţii din Cluj* (1920): 35–53.

from the other Romanian universities and significantly help to give it its own cachet.[21]

Another specific characteristic was the close relationship between the university and the city of Cluj, which engaged itself actively in the inauguration ceremonies, for instance by offering an afternoon cultural programme on 1 February 1920. The guests and local officials were invited to attend a performance at the Romanian Theatre and Opera House. The dramatic piece, titled *Poem of the Union*, portrayed the joy felt by Romanians after the recent historical events, which had enabled Transylvania to join the national state. Folkloric dances in traditional dress from all Romanian regions followed and were extremely appreciated by the audience. When the Royal Family, alongside the foreign delegates left the theatre, they were greeted by a public celebration, organized by the city's inhabitants and the university students. Many of the participants held burning torches, which created a very festive atmosphere, and after paying their respects, they marched through the main streets of Cluj, expressing their enthusiasm for the presence of the King and the opening festivities of the new Romanian university.

FIGURE 5.1 *Central building of Cluj University around 1926.*
REPRODUCED WITH KIND PERMISSION OF BABEŞ-BOLYAI UNIVERSITY HISTORICAL MUSEUM, CLUJ-NAPOCA

21 *Serbările pentru inaugurarea Universităţii din Cluj* (1920): 24–25. Puşcariu expressed similar ideas on other occasions and, as one of the founding fathers of the university, he

On the following day, 2 February 1920, the Royal Family visited all the buildings of the university (see figure 5.1), a military parade was organized on the main square, and after lunch the King received a delegation of Hungarian and Jewish inhabitants of Cluj at his residence.[22] The monarch and the Crown Prince rounded off the festivities with a special meeting with the students, in which they encouraged the youngsters to study and train in order to develop their country and to consolidate the newly acquired political union. In the late afternoon, the Royals and most of the foreign guests and national dignitaries left the city, while the students continued the celebration with a grand ball until late in the evening.

The inauguration festivities were regarded to be so important that the academic authorities decided to publish a series of special books to impress the event on the institutional and collective public memory. Two Romanian language publications were released, quite different in form and content, although with an almost identical title. The first of them, *Serbările inaugurării Universităţii româneşti din Cluj, 1–3 februarie 1920* [*Inauguration festivities of the Romanian University of Cluj*], offered a detailed description of each moment of the festivities in 62 beautifully presented pages. It included the full text of all the speeches delivered at the solemn inaugural ceremony in the Aula Magna, as well as the speech of the King in his meeting with the students. The general commitment of the authors with the Royal Family is shown by the closing sentence, stating that the Royals arrived back in Bucharest on 3 February 1920, at 8.45 a.m. The first half of the second book, *Serbările pentru inaugurarea Universităţii din Cluj, 31 ianuarie-2 februarie 1920* [*Festivities for the inauguration of the University of Cluj*], was largely a copy of the previous one. The second part of the 166 page book, however, included some interesting appendices: the congratulation messages received by the university on the occasion of its inauguration, a full list of the university professors and a generous selection of

actively supported this concept. Many of his successors at the rector's office carried it on. Amongst the chairs and scientific institutes founded exclusively within the Cluj University were the Institute of Speleology, the Museum of Romanian Language, and the Institute for the Study of Rabies. Moreover, until 1932, the Romanian University of Cluj was governed by special administrative rules (quite different from those applied at the other Romanian Universities of Bucharest and Iaşi), which maintained many institutional features inherited from Hungarian times – as for example the habit of electing a new rector at the start of each academic year.

22 *Serbările inaugurării universităţii româneşti din Cluj, 1–3 februarie 1920* [*Inauguration festivities of the Romanian University of Cluj, 1–3 February 1920*] (Bucharest: Imprimeriile Statului 1920): 58–59.

Romanian and foreign press articles, acclaiming the event and showing a pronounced sympathy for the Romanians.

However, it has to be observed that not everyone shared a positive attitude towards the celebrations in Cluj, although of course the editors of the books did not take into account these critical stances. The article of Ion Vinea, Romanian journalist and afterwards poet in the avant-garde movement, can function as an example in this respect. In an editorial published in the newspaper *Chemarea* [*The Call*], under the title "Kólosvàr" (sic),[23] Vinea used his customary polemic and acid tongue to criticize the inauguration of the Romanian university. He argued that the ceremonies were extremely artificial, as they took place in a city that was still utterly Hungarian. In his opinion, a cultural conquest could not be done merely by banquets and speeches.[24] Indeed, in 1920, the demographic statistics in Cluj showed 42.168 Hungarians, 10.638 Jews and only 29.644 Romanians in a total population of 85.509 inhabitants.[25] Consequently, despite his biting tone, Vinea was right in pointing to the fact that the new university was set in an intricate political, social and ethnic background and that Romanians had to be careful not to hurt feelings by displaying a rather arrogant behaviour.

Neglecting these warnings, a third book even appeared, dedicated to the 1920 academic festivities, being a French translation of the 166 paged Romanian book, with the title *Fêtes de l'inauguration de l'Université roumaine de Cluj. 31 janvier-2 février 1920* (see figure 5.2). The number of copies of this book is unknown, but assumingly it was sent as a present of thanks to most of the institutions and individual persons that had expressed their congratulations to the Romanian university of Cluj.

As an intermediate conclusion, it is obvious that the grandeur and complexity of the inauguration ceremonies marked a milestone and also established standards and traditions for the next academic anniversaries in Cluj, which often chose to repeat or reinvent the original model (sometimes by enhancing it, sometimes by replacing accents and even, for a while, by degrading it due to external factors). Through these academic celebrations, Cluj University also attempted successfully to impose itself as an institution

---

23 The Hungarian name of Cluj, Kolozsvár in the correct spelling, is used even up until today by the Hungarian inhabitants of the city.

24 Ion Vinea, "Kólosvàr", in: Ion Vinea, *Opere, vol. V. Publicistica 1920–1924*, edited by Elena Zaharia Filipaş (Bucharest: Academia Română 2003): 18–19.

25 Statistical data by Varga E. Árpád, *Erdély etnikai és felekezeti statisztikája (1850–1992)* [*Ethnic and religious statistics of Transylvania*], varga.adatbank.transindex.ro/?pg=3&action=etnik&id=5290 (date accessed 14/06/2014).

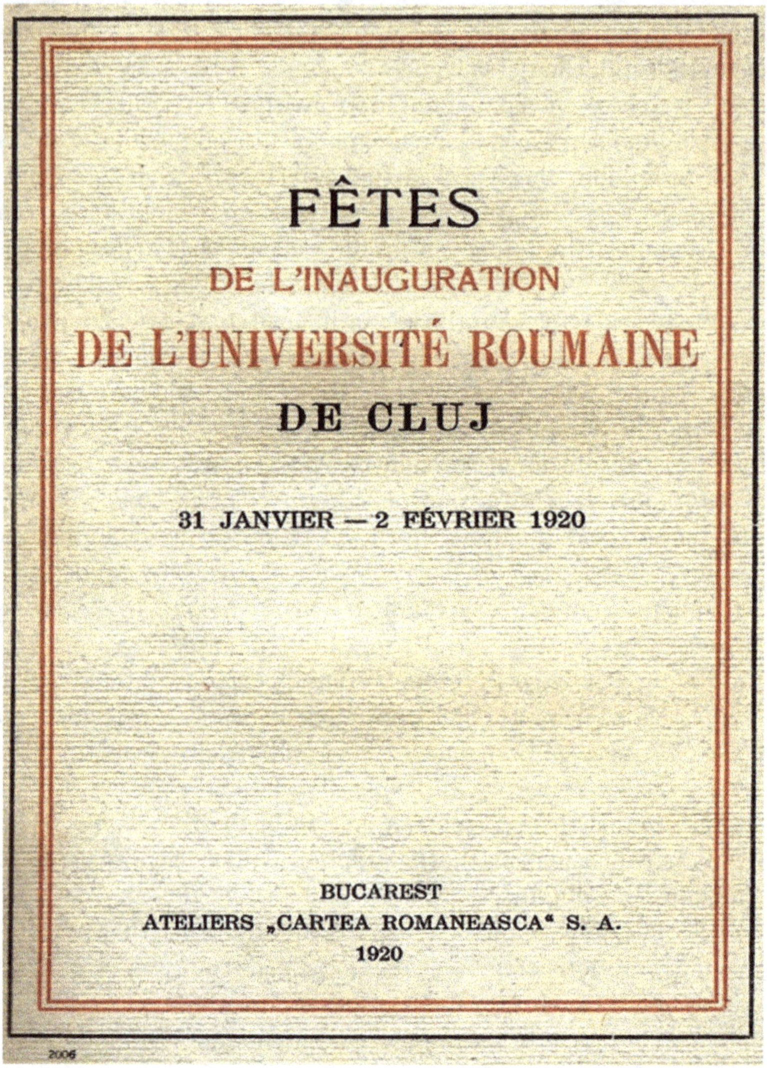

FIGURE 5.2 Fêtes de l'inauguration de l'Université roumaine de Cluj. 31 janvier-2 février 1920 (*Bucharest: Ateliers "Cartea Românească" S.A. 1920*): *cover.*
REPRODUCED WITH KIND PERMISSION OF "LUCIAN BLAGA" CENTRAL UNIVERSITY LIBRARY, CLUJ-NAPOCA

of reference for the other Romanian provincial universities. As a result, the Cluj inauguration ceremonies were used as a model at a similar event organized on 24–25 October 1920 in Cernăuţi/Czernowitz, Bukovina, where a second Romanian language university was officially opened at the end of the

First World War.[26] Nevertheless, one of the most significant differences between both festivities was the absence in Cernăuţi of diplomatic representatives from the Triple Entente and from the other newly created Central European countries,[27] a clear sign of the higher political and geo-strategic significance bestowed upon Transylvania by European governments.

The 10th Anniversary of the Romanian University of Cluj in 1930

The following celebration on the occasion of the 10th anniversary of the Romanian University of Cluj took place in a context of crisis, as the Great Depression of the 1930s was beginning to also manifest itself in Eastern and Central Europe. Politically and economically, Romania was under a lot of strain. However, this did not discourage Cluj University (which in 1927 had adopted the name "University King Ferdinand I", in honour of its founder and first royal patron[28]) from organizing a beautiful and rather sumptuous anniversary.

The then rector was the renowned scholar Emil Racoviţă, one of the founders of biospeleology. Being a prestigious and influential personality, he deployed assiduous efforts in order to create all the necessary conditions and actually carry out the celebrations. Racoviţă was also a firm believer that science can only flourish through worldwide cooperation and he himself had very close connections with many French universities. Racoviţă's correspondence with one of his French friends, the geographer Emmanuel de Martonne, unveils many of the details of the ceremonies. The preparations started in the summer of 1929, when Racoviţă was unanimously elected as the next rector "in order to [...] arrange and supervise the anniversary ceremonies of 2 February of the following year, to which the Maniu[29] government wants to give special

26 For a detailed description of the October 1920 academic festivities in Cernăuţi, see *Serbarea inaugurării Universităţii din Cernăuţi* [*Inauguration festivity of the Cernăuţi University*] (Bucharest: Imprimeria Statului 1920): 1–48.

27 The list of participants is published in *Serbarea inaugurării Universităţii din Cernăuţi* (1920): 6–7.

28 King Ferdinand I of Romania died in July 1927 and, as a sign of respect and admiration for his contribution to the academic life of Cluj, the Romanian university decided to officially adopt his name from October 1927 onwards.

29 Iuliu Maniu was a Romanian politician, of Transylvanian origin. He was the leader of the National Party of Transylvania and later of the National Peasants' Party of Romania. He was Prime Minister in three governments between 1928 and 1933.

importance". Already then, Racoviţă had some clear ideas about the general character of the festivities, wanting them to be international. To De Martonne he stated that it was impossible to celebrate the 10th anniversary of the Romanian university of Cluj without his presence, as "it is not about giving you another title [viz. that of Doctor Honoris Causa], but about reuniting all those who have made a contribution to the creation of Great Romania in general and the University of Cluj, inside this Great Romania, in particular."[30]

The date of the ceremonies was firstly postponed to May or early June 1930, when an unexpected political event once again altered the plans. On 6 June 1930, Carol II returned from his foreign exile, claimed the throne of Romania, and became King two days later. He immediately began to impose his authoritarian regime, largely based on a personality cult. In consequence of this political chaos, Cluj University postponed the festivities again, now to October 1930, combining the 10th anniversary celebration with the opening of the new academic year. The programme included all the classic ingredients (similar to those at the inauguration ceremony), but also many original elements. The main ceremony took place in the Aula Magna of the university on 20 October 1930, in the presence of several members of the Romanian government, about a dozen foreign guests (academic delegates and members of the diplomatic corps), and a delegation from the Royal Family (King Carol II, his mother – Queen Marie and Prince Nicholas, Carol's brother) (see figure 5.3).

For Carol II, it was his first official visit to a university after becoming King and thus a good opportunity to make public his attitude in educational, scientific and cultural matters. Consequently, in his congratulatory speech he addressed the issue carefully in order not to offend anyone. The King acknowledged the Transylvanian academic traditions, evoking the efforts deployed over many centuries by both the Hungarian and the Romanian communities in order to have a university in Cluj.[31] After praising the accomplishments of

30 Gheorghe Racoviţă and Ana-Maria Stan (eds.), *Memoria documentelor. Cooperarea franco-română la Universitatea din Cluj, oglindită în arhiva lui Emil Racoviţă* [*The memory of documents. French-Romanian cooperation at the University of Cluj reflected in the archives of Emil Racoviţă*] (Cluj: PUC 2007):153. Letter of Emil Racoviţă to Emmanuel de Martonne (06/06/1929).

31 King Carol II started his address by recalling the medieval college founded in 1581 in Cluj by Prince Stephen Báthory and continued with other significant moments over the following centuries. Although he did not explicitly mention 1872 as its founding year, he did evoke "the memorable Imperial Decree that created in Cluj the third Hungarian university", thereby implicitly recognising the existence of a Hungarian academic tradition in this city, before the creation of the Romanian university.

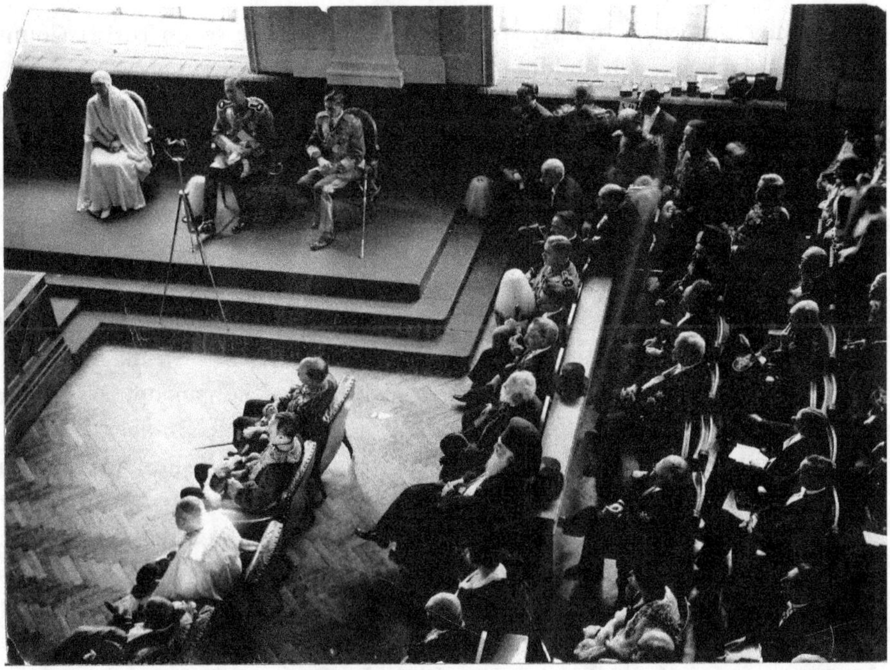

FIGURE 5.3 *On 20 October 1930, the Romanian Royal family (Queen Marie, King Carol II,* HRH *Prince Nicholas – on the podium) attends, alongside professors and foreign guests, the solemn ceremony in the Aula Magna to celebrate the 10th anniversary of the foundation of the Romanian University.*
REPRODUCED WITH KIND PERMISSION OF BABEŞ-BOLYAI UNIVERSITY HISTORICAL MUSEUM, CLUJ-NAPOCA

the young Romanian institution, Carol II announced the establishment of a scientific institute bearing his name.[32] However, unlike his predecessor in 1920, the new monarch did not provide a budget, nor did he fix a specific research area for the new institute, instead he asked for the help of the academic community in this endeavour.

The anniversary ceremony continued with a very special event – Queen Marie and seven other foreign personalities were awarded the title of Doctor Honoris Causa by the Romanian University of Cluj, which was a first in the life of this institution. The list of laureates included Metodie Zavoral (the Abbot of

32 Ioachim Crăciun (ed.), *Serbările jubiliare ale Universităţii din Cluj la împlinirea primului deceniu 1920–1930 şi activitatea ştiinţifică în primul deceniu 1920–1930* [*Jubilee celebrations of the University of Cluj on reaching the first decade 1920–1930 and its scientific activity in the first decade 1920–1930*] (Cluj: Cartea Românească 1930–1933): 9–10.

the Strahov Monastery in Prague), Matteo Bartoli (professor in Turin), Count de Saint Aulaire (French diplomat), Henry Roger and Emmanuel de Martonne (both of them professors in Paris), Henry Wickham Steed and Robert William Seton-Watson (well-known British scholars). Each of them had clear connections to Romania, having built up a profound knowledge of the country in previous years, studying its characteristics and some even pleading the Transylvanian cause long before the First World War. Therefore, they were honoured as friends of Great Romania and also as scientists who helped to introduce Romanian professors into the international academic networks.[33] Not all the seven foreign laureates were actually present in the Aula Magna of Cluj University – the British professors were most conspicuous by their absence. Those who were able to attend gave eloquent acceptance speeches, expressing their great admiration for the performances of the Romanian Alma Mater. De Martonne summarized it very nicely: "Besides giving excellent teaching, which is a huge task in itself, you have developed research with a brilliance that does honour to Romanian science. New chairs and laboratories have been created [...], all the publications of Cluj University can really stand up to comparison with those of the oldest universities. The fame of your professors has gone beyond your borders."[34] The ceremony in the Aula Magna was closed by King Carol II, who personally awarded some decorations to the most prominent university professors.

Whereas the first part of the festivities had a pronounced scholarly, intellectual and also international character, the afternoon of 20 October 1930, focussed on the place of the university within the city. The participants at the anniversary spent their afternoon at Cluj Sports Stadium, where approximately 10.000 inhabitants joined them. The entire audience watched a parade of different student and youth societies who also demonstrated their sports skills.[35] The leading role was given to the *Şoimii Carpaţilor* (Carpathian Hawks). This association was founded in the spring of 1930 by Iuliu Haţieganu, professor at the Faculty of Medicine of Cluj University, and focussed on educating worthy Romanian citizens by combining sports, folk dances and theoretical lectures on different subjects. It included mostly students, but also peasants or young

33 Crăciun (ed.), *Serbările jubiliare ale Universităţii din Cluj* (1930–1933): 33–65.

34 Crăciun (ed.), *Serbările jubiliare ale Universităţii din Cluj* (1930–1933): 18.

35 A short film dedicated to the 1930 academic festivities taking place in Cluj, which shows images of King Carol II and Queen Marie of Romania together with the rector of the university, Emil Racoviţă, as well as from the sports parade at the stadium, can be seen online at www.britishpathe.com/video/king-carol-of-romania-aka-king-caral-of-roumania (date accessed 15/06/2014).

pupils.[36] The Carpathian Hawks were partially modelled upon the Scouts, while also hinting at nationalistic, bio-political, and even eugenic principles (for example the selection of healthy, fit individuals, prepared and willing to serve the state at any moment), which had become part of the public debate in interwar Romania.[37] It was an impressive moment in the stadium, as it also hinted towards the rising preference of King Carol II for a specific model of the society, heavily influenced by Fascist Italy. After the sports parade, two football matches were played between local and national student teams and a cup was handed to the winners. Late in the evening, the Royal Family and the other guests attended a dramatic performance written by a local author at the Romanian Theatre and Opera House, just as it had been in the programme of the 1920 celebrations.

The following day, 21 October 1930, everyone's attention was fixed on the opening of the new academic year. The investiture of the new academic administration was definitely more solemn and full of splendour in the presence of the Royals. The ceremony continued with a special inaugural lecture, given not by the new rector, as was customary, but by one of the most distinguished university professors in Cluj. The honour of speaking in front of the King and the invited foreign academics fell upon Sextil Puşcariu, a leading linguist, who had also been the first rector of the Romanian university and who had played a central role in its inauguration.

Two other events of the 10th anniversary are worth mentioning, as they emphasize the scientific and cultural character of these festivities. A private collector had donated his famous Romanian paintings collection to the university, and so the academic authorities inaugurated the first pinacotheca in Cluj on 21 October 1930. On the same afternoon, a retrospective exhibition was opened at the university library. In two rooms, the Royals and the larger public could see all the scientific books and papers written by every member of the academic community of Cluj since 1920 displayed, as well as the experimental

36 Teodor Spârchez, "Şoimii Carpaţilor [Carpathian Hawks]", *Transilvania* 65 (1934), no. 1: 26–30; Bogdan Popa, "Cultura fizică: o ideologie a modernizării [Physical culture: an ideology of modernization]", *Cuvântul* 15 (2009), no. 6/384, http://www.cuvantul.ro/articol/?artID=5&nr=377 (date accessed 15/06/2014).

37 Marius Turda, "Controlling the national body: Ideas of national purification in Romania 1918–1944", in: Christian Promitzer, Sevasti Trubeta and Marius Turda (eds.), *Health, hygiene and Eugenics in Southeastern Europe to 1945* (Budapest: CEU 2011): 325–351; Marius Turda, "The nation as object: race, blood and biopolitics in interwar Romania", *Slavic Review* 66 (2007), no. 3: 413–441 and Maria Bucur, *Eugenics and modernization in interwar Romania* (Pittsburgh: University Press 2002): 19–233.

devices that they had invented or improved.[38] The exhibition ran until May 1931, and it could be considered the first attempt at a scientific and historical university museum in Cluj. Its ambition was to show the relevance of the scientific work done by Romanian professors and students and to compare it with the activities of the former Hungarian university, as many voices had doubted the capacity of Romanians to develop and maintain an academic life in Transylvania. The exhibition also provided the perfect opportunity to develop a thorough bibliographical overview, later published in book form, which catalogued all the scientific contributions of the teaching or research staff of the Romanian University of Cluj. In total it included 5761 titles published between 1920 and 1930, 1837 of which were written in a foreign language (meaning other than Romanian).[39]

The academic authorities also wanted to publish a special book dedicated to the 10th anniversary, as was done by their predecessors on the occasion of the inauguration. However, financial difficulties hindered this project, and although the material was finally gathered together in 1933, it was only printed and never actually produced in book form before the beginning of the Second World War. Only in 1949 were five copies of this very special edition finally published, creating a 528 pages volume. Nowadays, this book, titled *Serbările jubiliare ale Universităţii din Cluj la împlinirea primului deceniu 1920–1930 şi activitatea ştiinţifică în primul deceniu 1920–1930* [*Jubilee celebrations of the University of Cluj on reaching the first decade 1920–1930 and its scientific activity in the first decade 1920–1930*] is a rarity.

Despite the complex financial and political context, the 1930 anniversary of Cluj University can be considered a great success. The institution was at the centre of all the excitement, and by way of impressive festivities it emphasized science and culture much more than in the 1920 inauguration. The university acted as an autonomous and prestigious establishment, with a defined presence and role in the public scene of Cluj, while its professors and students were perceived as an elite group, with both civic and scientific responsibilities.

Over the next decades, a similar atmosphere was conspicuous by its absence. The 20th anniversary of the Romanian university was an atypical one: there were no grandiose ceremonies involving the inhabitants of Cluj or the local and national official establishment. Due to the general political tensions announcing the impending Second World War and to the rise of revisionism,

38 Crăciun (ed.), *Serbările jubiliare ale Universităţii din Cluj* (1930–1933): 128–129 and 137.

39 Ioachim Crăciun, *Activitatea ştiinţifică la Universitatea Regele Ferdinand I din Cluj în primul deceniu 1920–1930* [*The scientific activity at the King Ferdinand I University of Cluj in the first decade 1920–1930*] (Cluj: Cartea Românească 1935): 11–16.

the festivities were short and sober. The academic authorities partially broke the short established tradition, by choosing to celebrate not in 1940, but on 12 May 1939 – exactly 20 years since the Romanian government assumed control of the existing academic patrimony of the Hungarian university in Cluj. Onisifor Ghibu, professor in pedagogy, was in charge of organizing the anniversary celebrations. They included a religious service to commemorate the founding fathers of the Romanian university, namely those professors who took over the university buildings in 1919–1920 and later organised Romanian academic life, followed by a solemn meeting in the Aula Magna. The day finished with a ceremonial dinner at the university hall, in which only the professors, and not the students, took part.[40] The 20th anniversary was also an appropriate moment to reflect upon and write again about the accomplishments of the Romanian university during its short existence; and consequently, Ghibu described the evolution of the university in a booklet of 64 pages.[41]

Academic Anniversaries in the Communist Era – from King Ferdinand I University of Cluj to Babeş-Bolyai University

In the second half of the twentieth century, the profile and significance of the academic jubilees in Cluj were deeply altered. For almost fifty years, the academic life in Romania, and that of Cluj in particular, faced a great deal of hardship and underwent a complete makeover, generated by military and geopolitical upheavals. During the Second World War, the Transylvanian university did not suspend its activities, but had to move into other cities (viz. Sibiu and Timişoara),[42] as a result of the new border between Hungary and Romania, which Hitler and Mussolini had agreed upon in late August 1940.[43]

40 *Anuarul Universităţii Regele Ferdinand I din Cluj, 1938–1939* [*Yearbook of the King Ferdinand I University of Cluj, 1938–1939*] (Cluj: Cartea Românească 1940): 44–45.

41 Onisifor Ghibu, *La a douăzecea aniversare a Universităţii Daciei Superioare* [*On the twentieth anniversary of the University of Superior Dacia*] (Cluj: Institutul de Arte grafice 1939).

42 Puşcaş, *Universitate, societate, modernizare* (2003): 539–567; Marcela Sălăgean, "Universitatea din Cluj între 1919–1944 [The University of Cluj between 1919–1944]", in: Ovidiu Ghitta (ed.), *Istoria Universităţii Babeş-Bolyai* (Cluj: Mega 2012): 175–183.

43 The division of Transylvania between Hungary and Romania was decided on 30 August 1940, in Vienna, in a meeting between Hitler and Mussolini. The representatives of the Romanian and Hungarian governments were merely called to acknowledge this decision, which awarded the northern part of Transylvania (43.492 km^2 and approximately 2.5 million inhabitants, including the city of Cluj) to Hungary. See Corneliu Grad, *Al doilea arbitraj de la Viena* [*The second Vienna arbitration*] (Iaşi: Institutul European 1998): 37–58 and 100–124.

At the end of the War, the new communist regime, which gripped the country in an iron fist, gradually imposed its institutional and ideological vision upon the whole education system. Some of the fundamental changes experienced by the academic world after the introduction of communism in Romania are summed up very accurately in a fragment written by one of the university rectors of Cluj:

> Higher education was integrated into the scientific revolution in full swing and this became apparent in three essential aspects: a) the modification of the ideological content of the educational process and its substantiation by the scientific bases of Marxist-Leninist philosophy; b) the change of the relationship between culture and the broad mass of working people, clearing the way for the sons of the working class to attend institutes of education; and c) the education of the students in a spirit of patriotism and socialist internationalism, providing them at the same time with many-sided instruction on the level of advanced contemporary science.[44]

There were numerous other methods used by the communists in order to limit and transform academic autonomy according to their political vision. The implementation of communist values and ideas entailed massive waves of purges, especially in the late 1940s and throughout the 1950s, which resulted in a drastic selection of the teaching staff.[45] Only those who complied with the requirements of the new social and political Romanian authorities and proved their trustworthiness towards the Communist Party managed to keep their jobs within the academic system. Ideology thoroughly penetrated teaching and research, being constantly taken into account in all scientific publications. One of the typical examples is the book *Învăţătura lui I.V. Stalin cu privire la ştiinţa limbii şi sarcinile lingviştilor din Republica Populară Română* [*I.V. Stalin's teachings regarding the science of language and the tasks of linguists of the Romanian Popular Republic*],[46] written by professor Emil Petrovici, the first

44 Ştefan Pascu, "The History of the University of Cluj", in: *Babeş-Bolyai University* (Cluj: Întreprinderea Poligrafică 1971): 41–42.

45 For an in-depth analysis of the purges that took place at Cluj University, see Minodora Maria Cioban-Someşanu, *Epurări în universitatea clujeană 1944–1958* [*Purges in Cluj university 1944–1958*] (Bucharest: Academia Română, INST 2010): 18–28 and 48–183 and Mihai Teodor Nicoară, "Universitatea din Cluj în anii 1945–1959 [The University of Cluj between 1945 and 1959]", in: Ghitta (ed.), *Istoria Universităţii Babeş-Bolyai* (2012): 201–214.

46 Emil Petrovici published this book in 1951, under the auspices of the Academy of the Romanian Popular Republic.

communist rector of the University of Cluj, appointed by the new regime in 1945.

A new law of education adopted in 1948, followed by many other regulations, gave birth to a whole new internal structure for Cluj University in terms of faculties, chairs, departments and research institutes. Most importantly, the university was no longer the only academic institution in the city, as the faculty of medicine was ripped away from its old premises and transformed into a separate academic institute. Moreover, an agricultural academy and a polytechnic institute were founded, as well as a music conservatory and a conservatory of fine arts.

Practically and symbolically, Cluj University was somewhat "downgraded", because due to all these measures it came to be regarded by the local and national political authorities as just one academic establishment among a series of equals, in accordance with one of the basic principles of the communist system – egalitarianism. Furthermore, in 1945 the government decided to create a completely separate university for the Hungarian minority of Transylvania, which after a while adopted the name 'Bolyai' (after a family of Hungarian mathematicians, father and son, from the nineteenth century). The new university had several faculties and an entirely Hungarian teaching staff and administration.[47]

And the reforms continued. In the communist republic that Romania had become since January 1948,[48] any reference to the Royal Family was strictly forbidden, so that an alteration of the official name of Cluj University was required. The institution no longer honoured its prestigious founder, King Ferdinand I, but adopted instead the name of the physician and biologist Victor Babeş.[49] Born in Vienna in 1854, while the Austro-Hungarian Empire was still in existence, in a family of Romanian origin,

---

47 The Hungarian Uuniversity was founded by law on 28 May 1945. This act stipulated that the Romanian university of Cluj had to support and collaborate with this new institution, by accepting that Hungarian students use its laboratories and other practical facilities.

48 The last King of Romania, King Michael I, was forced to abdicate under Romanian and Soviet communist pressure on 30 December 1947 whereupon Romania immediately became a people's republic. The first Romanian communist constitution was adopted on 13 April 1948.

49 The change of name was discussed by the Cluj University Senate on its meeting of 10 January 1948 and officially acknowledged on 26 May 1948, when the Romanian communist government adopted a decree in this respect. Besides Babeş' name, there were four other names of Romanian personalities taken into account. Babeş-Bolyai University Archives: *Proces-verbal al Senatului universitar nr. 145* [*Minutes of the University Senate, no. 145*] (10/01/1948).

Babeş enjoyed international scientific fame during his lifetime. His name was chosen to patronize the university not only because of his biographical and professional qualities, but also because of his "democratic opinions", which were at the time positively appreciated and sanctioned by Soviet personalities.[50] The ban on the Romanian Royal Family, imposed by the communists, also affected the way in which future academic festivities were organised in Cluj.[51] The royal visits in 1920 and 1930 and the anniversary celebrations in which they participated were erased from the institutional memory of the university. The reference point for academic celebrations became the year 1919, when the Romanian university was founded and began to function, but even in that regard the essential role of King Ferdinand I was deliberately and completely ignored.

In 1959, there was no time or place for celebrating 40 years of activity – the subject of the day being instead the creation of the 'Babeş-Bolyai' University, by the unification of the Romanian 'Babeş' University of Cluj with its Hungarian counterpart – the 'Bolyai' University.[52] This decision was imposed by the government, marking not only a significant new era in the academic life of Cluj, but also a profound and gradual shift towards a more national approach of communism in Romania, obviously implying a change in the position of the communist authorities towards the country's minorities, especially the Hungarians. A few years before, in the autumn of 1956, a deep conflict had opposed the USSR and Hungary, culminating in the anticommunist Budapest revolt[53] (one of the first serious cracks in the unity of the Eastern bloc and an illustration of the fierceness of the Cold War period). The impact in neighbouring Romania was profound and manifested itself on many levels. First, the Hungarian minority of Transylvania, and in particular Hungarian intellectuals and academics, publicly supported the Budapest events. Furthermore, students from many academic cities in Romania, both Hungarian and Romanian, reacted energetically and manifested anticommunist opinions,

50 The deans and professors who took part in the senate meeting of 10 January 1948 repeatedly mention an article in favour of Victor Babeş, published at the time by a Soviet scientist in the *Analele Româno-Sovietice* 2 (1947), no. 7: 48–60.

51 The jubilees during the communist period have been reconstituted mostly on the basis of press and archival sources, since the number of books and articles addressing this subject is deficient.

52 For a description of the unification process, see Vasile Vese, "Universitatea Babeş-Bolyai în perioada regimului comunist 1959–1989 [The Babeş-Bolyai University during communism 1959–1989]", in: Ghitta (ed.), *Istoria Universităţii Babeş-Bolyai* (2012): 261–267.

53 Béla Király (ed.), *The first war between socialist states: The Hungarian revolution of 1956 and its impact* (Colorado: Boulder 1984).

which, however, were rapidly and severely repressed by the Romanian communist authorities.[54] Consequently, the 1959 unification of the Romanian and the Hungarian universities in Cluj and the subsequent establishment of the Babeş-Bolyai University was one of the punitive and coercive measures made in order to avoid any future deviations from communist ideology in Romania, particularly in Transylvania.

The year 1969 marked the semi-centenary of Cluj University under the Romanian administration, and although there was no separate, official ceremony to commemorate the event, it was duly acknowledged by the state authorities and the academic community. The opening of the academic year 1969–1970 was organised in grandiose style, with much more pomp than usual. Unlike on previous occasions, the solemn ceremony was not organised inside the university, in the Aula Magna, but at a new venue – the brand-new Students' Cultural House, where the main hall could receive almost 1000 participants. On 1 October 1969, students gathered here in large numbers, in order to receive the guests – national and local Communist Party officials, as well as the leading figures of all the academic institutions in Cluj. The most important dignitary was comrade Paul Niculescu-Mizil, secretary of the Central Committee of the Romanian Communist Party (P.C.R.). He was joined by Aurel Duca, the first secretary of the Cluj P.C.R. Committee; Traian Pop, the Vice-Minister of Education; rectors, professors and other highly placed persons.

In their addresses, Niculescu-Mizil and Duca mentioned the university's 50th anniversary, yet the stress was put on other issues. Both of them focussed on the differences between the communist regime and the previous political and social conditions, praising the present situation in heavily charged ideological wordings. "Our socialist revolution, the Communist Party and its fair national policies, of a Marxist-Leninist character, have managed to remove completely and permanently social as well as national discriminations, and limitations, enabling a large and complete access for worker's sons, regardless of their nationality, to education, science and culture," claimed Niculescu-Mizil.[55] Repeated references were made to the 10th Congress of the Romanian Communist Party and its directives, as well as to the recent law of education of 1969. The Party and academic leaders constantly stressed that the students, be they Romanians, Hungarians or Germans, needed to educate themselves in

54 Johanna Granville, "If hope is sin, then we are all guilty: Romanian Students' Reactions to the Hungarian Revolution and Soviet Intervention, 1956–1958", *Carl Beck Papers in Russian and East European studies* (2008), no. 1905: 1–76.

55 Discourse of Paul Niculescu-Mizil, published in the newspaper *Făclia* 24 (02/10/1969), no. 7125: 3.

order to "support the development of the society towards socialism". It was clear to them that an advanced scientific and cultural training could not be dissociated from a political one.

Apart from these long, tedious speeches, the 1969 anniversary did not distinguish itself much from the opening of previous academic years. At the end of the solemn session, groups of students delighted the audience with a music and theatre show, in the spirit of communist art. The assembly also decided to send a telegram to the Romanian president, Nicolae Ceauşescu, in which the students and the professors of Cluj vowed to make every effort to translate the Communist Party's guidelines into practice. Finally, as being the politician of the highest rank, comrade Niculescu-Mizil visited several buildings of the university, paying special attention to the site of the new student campus, which was designed to provide accommodation for most of the youngsters coming to Cluj to enrol in the university.[56] Infrastructure investments were a favourite topic of the communist regime, in its bid to transform Romania into a "multilaterally developed society".

Shortly afterwards, in 1972, another historical moment marked the academic life in Cluj, as 100 years had passed since the foundation of the first university in the city – the 'Ferenc József' Hungarian University. Even though the Romanian authorities had administered the institution throughout its different forms, only from 1919 onwards, did the local academic community internally acknowledge the importance of this landmark. The centenary of uninterrupted university teaching and research in Cluj, always in the same buildings, made them proud and aware of a long-standing tradition. Even the communist authorities in Bucharest indirectly agreed to celebrate this significant anniversary. On 2 October 1972, Ceauşescu himself paid a visit to Cluj and opened the new academic year with a mass ceremony. The dimension of this solemn meeting was bigger than ever, as thousands and thousands of students and workers were expected to go and greet the Romanian President at the city's Sports Hall. The local and national press outlined the fact that in 1972, due to the meticulous guidance of the Communist Party, Cluj had become the second academic centre in Romania. More than 22.000 students (five times more than in 1938) had studied in Cluj in that year, divided over 27 faculties, in a total number of six state universities – the Babeş-Bolyai University being the largest of them.[57]

In his very detailed speech, later published as a separate brochure, Ceauşescu however did not mention the academic centenary at all. Instead, he endlessly

56 *Făclia* 24 (02/10/1969), no. 7125: 1 and 3–4.

57 *Făclia* 28 (03/10/1972), no. 8053.

repeated propagandistic ideas about the role of the university professors and the tasks of the students. While the teaching staff was advised to strive to integrate the university courses with research and economic production (and thus keeping up with the world's scientific and technological development), the students were urged to play an active part in the creation of a "multilaterally developed society" – i.e. the communist society.[58] The only official representative who explicitly addressed the issue of the centenary during the solemn ceremony was the rector of Babeş-Bolyai University, Ştefan Pascu, but doing so in the presence of the Party leaders meant that they had approved it. Rector Pascu, who was also a famous historian, briefly outlined the old academic traditions of Cluj and, after pointing out that 1972 was the jubilee of the centenary, he continued that all the students and professors – Romanians, Hungarians and Germans alike – had to be grateful for the presence amongst them of Nicolae Ceauşescu, "the most beloved and esteemed Leader".[59]

One can easily observe that the public ceremony of 1972 did not exclusively focus on the centenary, as would have been customary, and became instead an occasion for praising the President, his wise management and his communist vision of Romania. However, that did not prevent the Babeş-Bolyai University from marking the occasion in different and more significant ways. For instance, Pascu, only just re-elected as rector, managed to publish a short history of the university on the occasion of its centenary. Under the neutral title Cluj University, Pascu retraced in fact the whole institutional academic life of Cluj, descending back to the Middle Ages. He dedicated special chapters to the Hungarian University of Cluj, as well as to the 1919–1945 period, but without a single mention of the role of the Royal Family in university life in the latter period.[60] The book was printed, in an identical form, in three separate editions – one Romanian, one Hungarian and one German –, thus acknowledging Transylvanian multiculturalism.

Recently accessible documents regarding the 60 year anniversary of the Babeş-Bolyai University also shed new light upon the profound alterations that characterised all the academic celebrations of the communist era. The 1979–1980 celebrations had an ambitious programme that was proposed by the University Senate, but that could only be executed after having received a favourable judgement from the communist authorities. The approval of

58 Nicolae Ceauşescu, *Cuvântare la festivităţile organizate la Cluj cu prilejul deschiderii noului an universitar* [*Speech at the festivities organised in Cluj for the opening of the new academic year*] (Bucharest: Ed. Politică 1972): 15–18.

59 *Făclia*, 24 (02/10/1969), no. 7125.

60 Ştefan Pascu, *Universitatea din Cluj* [*The University of Cluj*] (Cluj: Dacia 1972).

the Romanian Communist Party was essential and non-negotiable for each event that wished to mark a milestone in Cluj university life and only that which passed political censorship was allowed to happen. After obtaining the authorization of local Party authorities, the project was sent to the Minister of Education and Culture, to receive the final seal of ideological approval.[61]

In contrast to the previous festivities, in 1979 there was a partial reconciliation with the interwar tradition of Cluj University, as well as indications of a more relaxed atmosphere and a better cooperation between the Hungarian and Romanian professors and students. For the first time, some of the university rooms were named after outstanding local scholars – most of them having taught in the 1920s and 1930s, even if King Ferdinand's contribution to academic life in Cluj still remained unacknowledged. Also, in the university's main building, statues of Victor Babeş and János Bolyai were unveiled, side by side, in an attempt to mark the good relationship between the two academic communities that coexisted in Cluj. Significantly, these sculptures were commissioned and offered to the university by the Cluj Departmental Committee of Socialist Education and Culture, reflecting the Party's view of the majority-minority ethnic relationships. Besides scientific meetings in each academic discipline, the anniversary programme also included matches between the university's football and basketball teams and similar academic teams from Bulgaria and Poland.[62]

All in all, academic anniversaries in the communist period were interesting events. The atmosphere was deeply ideological and the university had become a mere tool that accurately reflected and applied the fluctuating conceptions and policies of the communist authorities. Now, its main task was to contribute, even through the festive moments, to the creation of the "new man", so highly praised in Marxist-Leninist texts and Ceauşescu's speeches. Consequently, the attention of the Party officials focussed on the students, the future builders of the so-called "communist multilaterally developed society", while propaganda obstructed a real assessment of the strengths and weaknesses of the Babeş-Bolyai University as a whole, or a profound inquiry into its history. Progress in

61 Babeş-Bolyai University Archives: *Letters exchanged between the rector's office of the Babeş-Bolyai University and Romanian Communist Party local and central authorities regarding the 60 years celebration of the Romanian University of Cluj* (January–June 1979). The festivities would take place in December 1979.

62 Babeş-Bolyai University Archives: Rector's Office: *Programul pentru sărbătorirea a 60 de ani de la crearea Universităţii Româneşti din Cluj-Napoca (1919–1979)* [*Programme for celebrating 60 years since the creation of the Romanian University of Cluj 1919–1979*] (1979).

terms of teaching and research quality as well as infrastructure development was equally delayed, especially in the late 1970s and 1980s.

University Celebrations at Babeş-Bolyai University after the Fall of the Iron Curtain

The academic festivities in Cluj after the upheaval in December 1989 and the subsequent fall of communism, represent complex moments of recent history that call for a detailed examination. In the future, time will allow historians to distance themselves sufficiently in order to fully reveal the characteristics of these celebrations and objectively analyse their significance, as well as the specific role of each of the participants and the decisive factors.

From 1994 onwards, celebrations have been organised at Babeş-Bolyai University every five years, reflecting not only the accelerating changes in academic life, but also the ongoing and often difficult democratization of Romanian society as a whole. In what follows, the focus is on the 1994 and 2009 anniversaries, which marked respectively 75 and 90 years of existence of the Romanian University of Cluj. After an analysis of the available data, several common traits appear.

First and foremost, the Babeş-Bolyai University fully regained its academic autonomy, so gravely and profoundly affected during communism. Consequently, it once again became the main organizer and actor of the festivities and gradually (re)gained the status of *primus inter pares* amongst the other academic institutions in the city.

Secondly, there was a conscious and steady effort to revive old traditions and restore the specific spirit that had gradually been built up during the first decades of the Romanian university. Just as during the interwar years, the venues used for the festivities were within the walls of the university – either the Aula Magna, or other majestic and vast university rooms. Foreign guests (ambassadors, professors and researchers who had contacts with the academic community) as well as representatives of all the other Romanian universities were once again invited to join in the celebrations. They gladly accepted the invitation to come and openly examined the achievements and, sometimes, failures of the Babeş-Bolyai University. Many of the invited personalities were honoured the way they used to be, with honorary doctoral degrees, being another anniversary custom introduced in the interwar period. Jubilee concerts were organised on behalf of the university, in a similar way as the theatrical performances of the 1920s and 1930s. Most importantly, in-house historians were encouraged to re-evaluate and rewrite the story of Cluj University from 1919 onwards, highlighting as much as possible both the achievements and the less successful episodes. The first effects of this process did not take long to appear: even though in 1994, there was still no mention of Kings Ferdinand I and Carol II,[63] in the 2009 festivity, the Royal Family's contribution to local academic life was already fully and officially re-established within the institutional tradition of the university, erasing almost 50 years of unjust neglect and imposed censorship.[64] Also, brochures and special booklets, detailing the programme of the academic festivities or the most important speeches, started to be again in print, as in the first half of the twentieth century.

Thirdly, in the post-communist celebrations each faculty, department and chair within the Babeş-Bolyai University delved deeply into its own historical evolution and analysed its scientific and teaching activity. Many books and studies were published on these subjects, reviving the debate on the mission of the university and how it should develop in the next decades. In Newman's words, "reaching for the truth" and critically analyzing it were reinstated as the core values of Cluj academic life.

The element that differentiated the 1994 and 2009 celebrations from all their predecessors was a comprehensive questioning of the place and role of Romanian society, and implicitly the university, within the European Union. In 1994, the festivities took place between mid-May and 10 June, with highlights from 27 May onwards. During those days, each chair and university department organised meetings to discuss their scientific activity, to present the latest achievements and to highlight future research programmes, aimed at restoring the connections with the international scientific world, connections that had been mostly intermittent since 1948.[65] On 8 June 1994, the university hosted a

63 None of the personalities who spoke at the main anniversary ceremony on 10 June 1994, mentioned the Royal Family and its previous visits to Cluj in their discourses (neither the Babeş-Bolyai University Rector, Andrei Marga; nor the Minister of Education, Liviu Maior – who was a historian and had previously worked at Cluj University; nor the President of Romania, Ion Iliescu – who was the guest of the highest rank). See the brochure *Universitatea 'Babeş-Bolyai' Cluj-Napoca, 1919–1994. Alocuţiunile rostite la festivitatea aniversară* [*The Babeş-Bolyai University of Cluj-Napoca, 1919–1994. Speeches of the anniversary festivity*] (Cluj: Rector's Office of the Babeş-Bolyai University 1994): 1–31.

64 The process of reinstating the Romanian Royal family in the Cluj academic tradition had gradually started from 2001, when the Babeş-Bolyai University Historical Museum was inaugurated. In 2003, the former Romanian King Michael I, the nephew of King Ferdinand I, was awarded the titles of Honorary Professor and Senator Ad Honorem of the Babeş-Bolyai University. Also, in 2005, statues of King Ferdinand I and his wife, Queen Marie, were unveiled in the main building of the Babeş-Bolyai University.

65 For detailed description of the national and international conferences and debates taking place in each faculty, see the brochure *75 de ani de la înfiinţarea Universităţii Daciei*

general meeting of the Romanian rectors, discussing possible future reforms for an efficient Romanian university system. And this debate continued during the two following days, when Romanian and foreign specialists sat together in an international conference to reflect on the "role of universities for the construction of the new [united] Europe".

The main anniversary ceremony was saved for the last day of the celebrations, organized in the Aula Magna of the university, in the presence of Romanian governmental authorities, joined by representatives of Western universities and ambassadors of several influential countries of the European Union. The whole anniversary ended with a beautiful classical music concert.[66] Besides the variety of celebratory events, which ranged from scientific to cultural and even managerial, the 1994 anniversary marked a renewal of national and international academic connections of the Babeş-Bolyai University. No less than 115 foreign guests had visited the university during the festivities, as the then rector emphasized in his anniversary speech.[67] Obviously, many of these contacts later turned into interesting and lasting international partnerships, as well as teaching and research networks.

In 2009, with Romania being a full member of the European Union since 2007, the academic festivities had an even more distinct character. The entire academic year 2009–2010 was considered an anniversary year. Consequently specific events were organized at each and every level of the Babeş-Bolyai University. The time that had passed since the end of the communist regime had brought along another profound institutional transformation. From 1995 onwards, Babeş-Bolyai University was able to offer students the chance to study the whole academic curriculum – licentiate, master and PhD – in either Romanian, Hungarian or German, and most of the faculties were organised along these three lines of study. Also, the number of faculties experienced an unprecedented growth and development, reaching a total of 21 and offering specializations in a large variety of new scientific domains, such as environmental studies, political sciences, European studies, nanotechnology, ICT and the like. Each department and institute, from all the faculties and all the lines of study played its own part in the anniversary, by organising workshops, conferences and publishing interesting research. The rector's office also took part in the debates at several international conferences, with partners from

Superioare 1919–1994. Programul aniversării [*75 years since the foundation of the University of Dacia Superior 1919–1994. The anniversary programme*] (Cluj: Rector's Office of the Babeş-Bolyai University 1994): 12–127.

66 *75 de ani de la înfiinţarea Universităţii Daciei Superioare* (1994): 8–12.

67 The brochure *Universitatea 'Babeş-Bolyai' Cluj-Napoca, 1919–1994* (1994): 17.

academic networks such as the Santander Group or the Danube Rectors' Conference, discussing delicate issues such as the place of universities in a globalized world or how the economic world crisis could be dealt with. Concerts, sports competitions and quite a few scientific and artistic exhibitions completed the celebrations.[68] Last, but not least, a special new volume dedicated to the history of academic life in Cluj, starting back in 1581, was commissioned by the Rector's Office to the Faculty of History and has only recently been published.[69]

Conclusion

By analyzing the twists and turns of the academic ceremonies in Cluj during almost hundred years, we gain a fresh and interesting perspective on the institutional evolution of the university. One can easily observe the complexity of its history and reflect upon the challenges it had to face over the course of time. Looking back at the interwar years, it can easily be noted that the ceremonies of Cluj University incorporated all the customary elements and did not distinguish themselves too much from what happened in Western Europe.[70] A solemn ceremony in which congratulatory addresses were delivered and honorary doctoral degrees were awarded, in the presence of cultural, political and religious elites, formed the core event. Another aspect, visible in 1920 and even more pointedly in 1930, was the fact that these anniversaries provided the perfect opportunity to reinforce the relationships between the university and its students, as well as between the university and the regional and local population. The generic cultural aspect of the festivities was equally present.

In contrast, after 1948, the prevailing tendency in the ceremonies was to justify and support the efforts of the Communist Party in transforming Romanian society according to communist ideology. Politics deeply shaped the organization of jubilees, which evolved from Marxism-Leninism towards a more national form of communism, especially during the latter part of the 1980s. The university and its academic community had to serve and accurately reflect

68 The brochure UBB *90. Programul manifestărilor consacrate aniversării a 90 de ani de la întemeierea Universităţii româneşti din Cluj, 1919–2009* [*The events dedicated to the celebration of the 90th anniversary of the founding of the Romanian University of Cluj, 1919–2009. Programme*] (Cluj: Babeş-Bolyai University 2009).

69 Ghitta (ed.), *Istoria Universităţii Babeş-Bolyai* (2012).

70 For an analysis of the standard ingredients of university jubilees, see Bekker, "Jubiläen als orte universitärer Selbstdarstellung" (2008): 77–107.

the conceptions of the communist leaders, in particular those of Ceauşescu and his close entourage. Since the fall of the Iron Curtain and the return of Eastern Europe towards democracy, there is an ongoing effort to restore the academic tradition in Cluj and reflect upon its most constant, popular and valuable elements.

Two other significant elements need to be outlined. First, one should point out a certain fragility in choosing and imposing on the life of the Babeş-Bolyai University a single, distinctive date for the academic jubilees throughout the twentieth century and, consequently, a certain failure in establishing an uncontested, unifying academic tradition. This can be explained by the intricate institutional, local and national history. Secondly, we must reflect upon the constant rewriting/reshaping of university history in such fluctuating conditions. There is not one 'true', final Cluj university history, and this opens the field for future research. The titles of the books and brochures dedicated to Cluj University from 1920 onwards stand as a powerful testimonial to the numerous turns taken by university history writing in Transylvania (and in Cluj in particular) over the last century. In the first 50 years, the trend was to emphasize the idea of Roman (Latin) ancestors, sometimes even Dacian ones, in a bid to prove the continuity of Romanian cultural presence in the province and, implicitly, of their entitlement to a long and prestigious cultural heritage, that the 1919 new founded university took to another level. During the communist period, university history writing reflected mostly another approach: the university, as an institution, became a place to put into practice the success of communist propaganda and to illustrate the cordial relationships between different ethnic groups. Since 1989, there is a continuous attempt to conciliate such diverse historiographical points of view and find common denominators.

Therefore, what happened from 1920 onwards can and should provide suggestions for future anniversaries, especially in 2019, when the Babeş-Bolyai University will celebrate a hundred years of existence since its foundation as a university of the Romanian state, in the context of Transylvanian multiculturalism. Writing the Babeş-Bolyai University History on the occasion of its centenary will be a challenge and a huge responsibility for its authors.

PART 2

University History Writing on the Occasion of a Jubilee

CHAPTER 6

1968 as a Turning Point in Trondheim's University History

Thomas Brandt

Writing the jubilee history of The Norwegian University of Science and Technology (NTNU) in Trondheim on the occasion of its centenary posed several challenges. It soon became clear that there was little consensus on what parts of the history ought to be treated and which period to be focussed on. The NTNU is a rather novel institution, resulting from the 1995 merger of a former regional university and the Norwegian Institute of Technology. In order to understand better the long-drawn differences over university identity in Trondheim, this chapter will first give a brief overview of the origins of the merged institutions in Trondheim. Secondly it will map out the history of the various crossroads that university leaders were faced with in the tense 1960s and 1970s. It will become clear that 1968 was a turning point in Trondheim's university history in general and in its search for an own identity in particular. In conclusion, some of the challenges of writing the history of such a particular kind of institution will be addressed.

In 2010 the Norwegian University of Science and Technology (NTNU) in Trondheim celebrated its centenary by organizing all kinds of scientific and cultural events, as well as the publication of a university history.[1] One of the challenges historians have to face when approaching the history of NTNU concerns the disagreements over the legitimacy of this rather novel institutional construction, a difficulty many historians of university history are faced with. As demonstrated by Emmanuelle Picard, for instance, the reconstruction of the French university landscape through a series of reforms has contributed to vast challenges in producing coherent institutional histories.[2] Formally, NTNU was established only in 1995 after more than thirty years of political struggle over the integration of Norway's most important engineering educational institution and the local university. While the institutional roots of the modern

1 Thomas Brandt and Ola Nordal, *Turbulens og tankekraft. Historien om NTNU [Turbulence and the Power of Thought. The History of the NTNU]* (Oslo: Pax forlag 2010).

2 Emmanuelle Picard, "Recovering the History of the French University", *Studium. Tijdschrift voor Wetenschaps- en Universiteitsgeschiedenis Revue de l'Histoire des Sciences et des Universités* 5 (2012), no. 3: 156–169.

 | DOI 10.1163/9789004265073_007

university construction lead back to The Royal Norwegian Society of Sciences and Letters, established in 1760, NTNU instead chose to celebrate the centennial anniversary of the Norwegian Institute of Technology, established in 1910.

This resulted in the engineering traditions being given the dominant position in the jubilee, despite the many other academic traditions present at NTNU. It also posed the following challenge for us as historians commissioned to write a history of this institution: even if the celebrations were to mainly focus on the engineering school traditions from 1910, how could we try to integrate the whole new university in a historical account? We chose to solve this, firstly by encompassing the 250 years of modern science in Trondheim since the establishment of the Royal Society. Secondly, instead of sweeping the many former differences of opinion under the carpet of celebration, we explicitly made the effort to map out the origins of the drawn-out struggle over university integration in Trondheim.

This chapter is, then, a case of "show and tell", in which an approach to the university history is demonstrated and some thoughts are offered on the political implications of writing history. In order to understand better the long-drawn out differences over university identity in Trondheim, this chapter will first give a brief overview of the origins of the merged institutions in Trondheim. Then it will map out the history of the various crossroads that university leaders were faced with in the tense 1960s and 1970s. In conclusion, some of the challenges of writing the history of such a particular kind of institution will be addressed.

From an Enlightened Learned Society to the Science Museum

In 1760, a learned society of sciences and letters was established in Trondheim, from where the foundation of the university's library and science museum can be traced. Trondheim was then a small town of some 7.000 inhabitants who earned a living by trading in fish, timber, copper ore and other forms of raw material for the benefit of more developed proto-industrial European economies.[3] The learned society was founded by a small group of natural historians and philosophers led by the newly appointed Bishop of Trondheim, Johan Ernst Gunnerus. Due to Gunnerus' entrepreneurial efforts, the society rapidly expanded to become a vital centre of Scandinavian Enlightenment, obtaining

---

3 Steinar Supphellen, "Preface", in: Håkon With Andersen, Brita Brenna, Magne Njåstad and Astrid Wale, *Aemula Lauri. The Royal Norwegian Society of Sciences and Letters, 17640–2010* (Sagamore Beach, MA: Science History Publications 2010): vii–viii.

the right to be called The Royal Society of Sciences and Letters in 1767. Gunnerus' extensive scientific network set in motion the circulation of large amounts of specimens, commodities, books and letters between Trondheim and other hubs of eighteenth-century science in Northern Europe. Most notable is the lasting correspondence with Carl Linnaeus in Uppsala, between 1761 and 1772.[4]

After Gunnerus' death in 1773, the Royal Society fell into decline and gradually became a patriotic rather than a scientific society, more concerned with awarding prizes for various practical and utilitarian achievements.[5] After the establishment of the *Kongelige Frederiks Universitet* in Norway's capital Christiania (Oslo) in 1811, there was even some discussion among the Society's members to incorporate the Society into the new university, as it was now in reality a scientific society without scientists.[6]

Still, the society managed to transform itself in order to survive, and built a library and a museum during the nineteenth century for the varied collections of specimens, objects and books that it had accumulated over the years. A more strategic and targeted way of collecting was started, in line with the upcoming scientific interest for systematization, collection and typology.[7] Gradually the science museum and library in Trondheim established itself as one of Norway's representatives in the international science museum movement, alongside the museums in Bergen (1825) and Tromsø (1840). While these two latter museums spearheaded the establishment of universities in their respective cities after the Second World War, the museum in Trondheim played only a minor role among several academic institutions debating upon the creation of a new university (see figure 6.1).

The Establishment of Higher Engineering Education in Trondheim

The most influential of the academic institutions in Trondheim involved in the discussions in the 1960s on the foundation of a university was *Norges Tekniske*

4 Leiv Amundsen (ed.), *Johan Ernst Gunnerus og Carl von Linné brevveksling 1761–1772* [*The Correspondance of Johan Ernst Gunnerus and Carl Linnaeus 1761–1772*] (Trondheim/Oslo/Bergen/Tromsø: Universitetsforlaget 1976).

5 Brita Brenna, "Prizes, patriots and scientific texts", in: Andersen a.o., *Aemula Lauri* (2010): 95–99.

6 Magne Njåstad, "A changing Society: 1805–1830", in: Andersen a.o., *Aemula Lauri* (2010): 103–128.

7 For this transformation, see Magne Njåstad, "New directions, 1850–1870" and Håkon With Andersen, "A new direction – away from a society of savants?", in: Andersen a.o., *Aemula*

FIGURE 6.1 *View of Trondheim in 1800. The building on the far-right from 1787 housed two of the city's early institutions of learning: the Cathedral school and the natural history collection of the Royal Norwegian Society of Sciences and Letters. Centre-left is the Nidaros Cathedral. Watercolour painting by Johan F.L. Dreier.*
REPRODUCED WITH KIND PERMISSION OF *NORDENFJELDSKE KUNSTINDUSTRIMUSEUM* (NATIONAL MUSEUM OF DECORATIVE ARTS), TRONDHEIM

Høiskole (NTH), which was founded as a technical college in 1910. The NTH soon developed into the central institution for technical science and higher engineering and architectural education in Norway.[8] The NTH was an educational institution that ranked alongside the university in the capital. Applicants were required to pass their *examen artium* (or an equivalent) in order to be admitted. A professors' council ran the college, with the rector as their *primus inter pares*.

The establishment of this first Norwegian institution of higher engineering and architectural education was the result of a long-drawn-out struggle

Lauri (2010): 180–182, 201–209, and John V. Pickstone, *Ways of Knowing: A New History of Science, Technology and Medicine* (Chicago/Manchester: University Press 2000): 73.

8 The translation of *Norges Tekniske Høiskole* into English is problematic. NTH is best likened to the German *Technische Hochschule*. After World War II, NTH's professors and officials often used the English name, Norwegian Institute of Technology, to reflect the orientation towards Anglo-American ideals. I prefer the term 'technical college' to indicate the linkage to a European tradition.

between politicians, industrialists, engineering organizations and the pundits of the press on what type of technical education Norway needed, and not least, on the physical location of such an institution.[9]

When, after many years of discussion, the Norwegian parliament finally decided in 1900 to establish a higher technical college, Trondheim was chosen, both because the city already had Norway's most advanced polytechnic school, and because of the general tendency towards decentralization as a motive in Norway's political discourse. In addition, Trondheim could offer an attractive endowment in terms of resources and facilities for the new college. Still, the decision to establish the nation's first university level education for engineers in Trondheim was heavily criticized by many of the leading intellectuals and industrialists in the capital. According to them, Kristiania should have been chosen, since it was only in this region of Norway that the synergies of engineering education, industry and university would have been possible.[10] Apart from a relatively large (by Norwegian standards) shipyard and some mechanical industry, there was little industrial activity to speak of in Trondheim.

Nevertheless, when the new technical college was inaugurated in September 1910, it happened in an atmosphere of optimism and pioneering spirit. The NTH consisted of 12 professors and 103 students, and was divided into seven faculties, viz. architecture, chemistry, civil engineering, electrical engineering, mechanical engineering and mining engineering, and a general science faculty providing courses in physics, mathematics and some elementary courses in social sciences. On the one hand the NTH was a blueprint of a German *Technische Hochschule,* especially with regard to the layout of the teaching laboratories and the incorporation of an institution for testing materials, but also in terms of the emphasis on proficiency within the engineering sciences. Graduates from the *tekniske høiskole* were to have a broad, yet theoretically advanced technical education.[11] Moreover, several of the professors were

9 Tore Jørgen Hanisch and Even Lange, *Vitenskap for industrien. NTH – En høyskole i utvikling gjennom 75 år* [*Science for the Industry. NTH – A College in 75 Years of Development*] (Oslo/Bergen/Stavanger/Trondheim: Universitetsforlaget 1985) and Brandt and Nordal, *Turbulens og tankekraft* (2010).

10 Hanisch and Lange, *Vitenskap for industrien* (1985): 31.

11 Wolfgang König, "Technical education and industrial performance in Germany: a triumph of heterogeneity", in Robert Fox and Anna Guagnini (eds.) *Education, Technology and Industrial Performance in Europe, 1850–1939* (Cambridge: University Press 1993): 68–81; Roland Wittje, "The Foundation of N.T.H. in 1910 in International Context", in Reinhard Siegmund-Schultze and Henrik Kragh Sørensen (eds.), *Perspectives on Scandinavian Science in the Early Twentieth Century* (Oslo: Novus Press 2006): 119.

recruited from Germany, and most of their Norwegian colleagues had graduated from German institutions. On the other hand the college in Trondheim was adapted to Norwegian conditions, both in terms of curriculum and in how the college was set up to educate engineers who would be intellectually self-sufficient. No technical matter should be unfamiliar to the NTH engineer, and he (it was only extremely rarely a 'she') should be able to lead and manage almost any kind of business involving technology.

The NTH was located on Gløshaugen, a hill overlooking the city centre, where the large main building dominated the horizon (see figure 6.2). With its two towers, it was said the facade deliberately resembled the Nidaros Cathedral just across the river. In reality, primarily financial concerns had decided on the architecture of the main building. It was the most inexpensive option of the shortlisted proposals. Yet, the coarsely cut granite stone covering the facade was clearly intended to symbolize Norway's mountains, from which the engineers at NTH should extract riches for the nation and harness hydro-power from the rivers. In order to do so, the students and their professors had several laboratories at their disposal, some of which were very much state-of-the-art teaching laboratories.

During the first decade of its existence, the activity at the NTH expanded substantially, resulting in 123 employees, 26 of them being professors, more

FIGURE 6.2 *One of the first aerial photos of the Norwegian Technical College in Trondheim, dated ca. 1933. The college facilities remained almost unchanged over the following two decades.*
REPRODUCED WITH KIND PERMISSION OF TRØNDELAG FOLK MUSEUM, SVERRESBORG

than 600 students and a number of new laboratories.[12] However, in consequence of the difficult economic situation in Norway after 1921, growth was halted and the college even experienced severe problems in the recruitment of both staff and students.[13]

The education of engineers and architects for Norway's industry and public services had become the main task of the NTH. In addition, the professors were expected to cater to the needs of industry as consultants and they were even encouraged to engage in entrepreneurial activities. Those who were interested in research activities had to opt for collaboration with industrial partners or else to resort to small-scale research activities in their spare time. Yet, the NTH was criticized by both students and the engineering community for being overly preoccupied with theoretical research.[14]

Between 1910 and 1940, the NTH gained a reputation for offering an attractive educational programme, and being a meritocratic institution that recruited some of the best and brightest students from the upper social strata. Trondheim also became known as having the most lively student culture in Norway, largely due to the *Studentersamfundet i Trondhjem* (often abbreviated to *Samfundet*). The association was established by the NTH students themselves shortly after the inauguration of the school in 1910. It became a central meeting place for both serious discussions and more sophomoric activities among students and professors alike.[15] By the time the Second World War broke out, students and teachers at the NTH had managed to create a vibrant academic culture. Also aided by the creation of a series of traditions like the professors' ceremonial robes, inaugural rituals and symbols, a strong NTH identity had emerged.

---

12 Brandt and Nordal, *Turbulens og tankekraft* (2010): 485–489.

13 Due to a failed monetary policy, Norway suffered hard and early from the 1920s international economic recession. See Sverre Knutsen "Finance and the development of Norwegian Capitalism: the case of Christiania Bank", in Susanna Fellmann a.o. (eds.) *Creating Nordic Capitalism. The Business History of a Competitive Periphery* (Basingstoke: Palgrave Macmillan 2008): 544.

14 The debate over this issue is summarized in Johan Herman Lie Vogt, "Om professorene ved den tekniske høiskole", *Teknisk Ugeblad* (18/02/1910): 73–74; See also Hanisch and Lange, *Vitenskap for industrien* (1985): 45 and Brandt and Nordal, *Turbulens og tankekraft* (2010): 108.

15 Jan Thomas Kobberrød, *Engasjement og Begerklang: Studentersamfundet i Trondhjem 1910–2010* [*The Students' Society in Trondheim 1910–2010*] (Trondheim: Tapir akademisk forlag 2010).

A Teachers' College based on *Volksbildung*

In 1922, the NTH was joined by an additional, quite different college for higher education. Once again, Trondheim had won the bid in the Norwegian parliament over the location of a national institution, in this case the Norwegian teachers' college (*Norges Lærerhøgskole i Trondheim,* NLHT). Since the late nineteenth century, the education of teachers in Norway had followed two separate tracks: university education predominantly led to teaching at the high-school level in the major towns and cities, while more basic teacher seminars more often resulted in teaching in rural primary schools. An ideological and cultural schism existed between these two institutions in the perception of how teachers should be trained.

In the early 1900s, a severe shortage of teachers at all levels became apparent and Norwegian educators heatedly debated how to remedy this. As in many other European countries, the discussion focussed on the question of whether teachers should be trained at the university or at a separate teachers' college.[16] Those in favour of a university education judged the seminar tradition to be biased towards esoteric, nationalistic and religious beliefs. On the other hand, those in favour of a separate teachers' college had doubts about the university's interest in the conditions within rural Norway. They were also promoters of an alternative modernity not based on technological, industrial and economic growth, but on ideals that could be found within the *Volksbildung* tradition, such as reform pedagogy and Christian, rural and national traditions.

The seminar movement emerged victorious from the debate with their claim for a separate teachers' college. However, at that moment, around 1920, the situation was not very favourable for establishing a new educational institution. Norway's economic situation looked difficult, and there was a call for frugality in public expenditure. Secondly, during this time the need for teachers at the level of lower secondary schools had become less acute, which resulted in the foundation of a very modest institution located in an old farmhouse, just outside the centre of Trondheim. The school had only three professors and 60 primary school teachers as their students. Pedagogy, Norwegian language, literature and history were the main subjects, although other courses such as theology, music, philosophy and modern languages were offered by local intellectuals as well. More substantially, the NLHT collaborated with the NTH in providing the necessary classes in mathematics and natural sciences.

16 Pieter Dhondt, "Teacher Training Inside or Outside the University: The Belgian Compromise (1815 to 1890)", *Paedagogica Historica. International Journal of the History of Education* 44 (2008), no. 5: 587–605.

Several of the NTH's most renowned professors in physics, chemistry and mathematics supported the new teachers' college, some of them even expressed their hope for a common academic culture in Trondheim.[17]

As a social group, the students at NLHT were quite different from the engineering students at NTH. Often they had already worked for some years as teachers in a rural community, and only a few of them had an education at university level. Mostly they relied on a small stipend in order to be able to complete the one-year course. Moreover, the short course was deemed insufficient by the authorities to allow for a final exam and in result the NLTH graduates had little to show for it in terms of diplomas. Their motivation for going to NLHT was often more a matter of conviction. Many of the NLHT students, as well as some of the professors, were active in the temperance movement. Even more belonged to the minority *nynorsk* movement which evolved during the Norwegian language conflict that had erupted in the first decades of the twentieth century.[18] This stood in stark contrast to the more indulgent NTH students, mainly coming from the upper echelons of Norway's urban society. Because of this difference in culture and scope of education, the students at NLHT were not allowed membership of the student's society in Trondheim until 1959.[19]

While the differences between the two institutions of higher education sometimes did become openly hostile, it would still be wrong to characterize the situation in Trondheim as only a crude example of the clash between the "two cultures" of literary intellectuals and scientists, as identified by Charles Percy Snow.[20] It would be more precise to describe the academic environment in Trondheim as an intersection of multiple scientific and scholarly cultures.

Post-War Urge for Modernization of the NTH

Both educational institutions in Trondheim, the NLHT and the NTH, had suffered a serious setback because of the German occupation between 1940 and

17 For example, NTH professor Richard Birkeland expressed his hope for a close collaboration in a committee report on the teachers' college: *Indstilling om Lærerhøiskole, avgit av en av Kirke- og undervisningsdepartementet nedsat komite* [*Report on a Teachers' College by a Committee Apointed by the Department of Ecclesiastics and Education*] (Kristiania: Chr. Gundersens boktrykkeri 1921).

18 Anders Kirkhusmo, *Akademi og Seminar. Norges lærerhøgskole 1922–1982* [*Academy and Seminar. The Norwegian Teachers' College 1922–1982*] (Trondheim: Tapir 1983): 64–71.

19 Kobberrød, *Engasjement og begerklang* (2010): 182.

20 Charles Percy Snow, *The two cultures and the scientific revolution* (Cambridge: University Press 1959): 4.

1945. The plans for further development that had been made before the war had already become obsolete. The Second World War had clearly demonstrated the importance of science and technology. Therefore Norway embarked on planning a post-war reconstruction of society, in which a stronger and gradually more instrumental emphasis was put on research and higher education. In the joint programme of the Norwegian post-war coalition government of 1945, scientific research in the service of industrial production was highlighted as an important way ahead for the nation.[21] The NTH had to struggle hard not to be sidelined in the process of restructuring the Norwegian research system. At the same time, the administration of the NTH was put under pressure by many students waiting, either to resume or to start their education after the war. The facilities were dilapidated and obsolete, and needed to be renovated and extended to meet the rapid increase in students and faculty. Between 1945 and 1965, the number of staff members increased from 217 to 921, while the number of students soared from 1.000 to 2.611. In the same period, the campus building area quadrupled in size.[22]

Yet the transition after the Second World War implied more than just a quantitative growth. New scientific fields connected to engineering and industry had emerged, such as electronic engineering and chemical engineering. Young professors, who had often studied or worked as a researcher at American or British institutions during or just after the war, urged for a radical transformation of the NTH's educational system, which had remained almost unchanged since 1910. They were backed by the interests of industry and the state to reconstruct the nation as fast as possible. This led to an intense process of reassessment for most of the NTH faculties' study programmes. The faculty of chemistry took the lead by the introduction of a study plan committee in June 1945, which was soon followed by several other committees. In addition, the influential Norwegian Association of Engineers (*Den Norske Ingeniørforening*, NIF) took the initiative to create a separate study plan committee, chaired by the general manager of the Aluminium Company Årdal og Sunndal Verk, Aage W. Owe. The agenda of this so-called Owe Committee was to present an overall design for the study course of architects and engineers at the NTH. The explicit ambition was to help to "modernize, rationalize and coordinate" the education at the NTH.[23]

---

21 The programme is reprinted in Jan-Erik Ebbestad Hansen (ed.), *Norsk tro og tanke bind 3* [*Norwegian Belief and Reason Vol. 3*] (Oslo: Tano Aschehoug 1998/2001): 231–239.

22 Brandt and Nordal, *Turbulens og tankekraft* (2010): 247.

23 *Studieplankomiteen for NTH, oppnevnt 1949 av NTH, NIF, NAL og Studentsamskipnaden i Trondheim. Innstilling av 25de oktober 1952* [*The Study Plan Committee of 1949, "The Owe Committee"*] (Trondheim: NTH-trykk 1952): 2.

Most likely, the Norwegian Association of Engineers had hoped that the committee would come up with suggestions on how non-technical subjects could be given more attention at the NTH.[24] The debate about how much space should be given to courses in humanities and social sciences was almost as old as the NTH itself. Already before the war, many of the school's graduates went on to careers as managers, and from time to time they lamented at their lack of general education in economics or industrial management. For instance, in 1934 the architect Ove Bang quipped that at the NTH the concern was to "educate small scientists".[25] On the initiative of some of the more intellectually inclined professors and students, a lecturer in philosophy was hired in 1917. Later on, in the 1920s, some short courses in social economy, law and sociology were introduced, though these were rather feeble attempts at providing some elements of general education at the NTH.

If the Norwegian engineers had indeed placed their hopes on the Owe Committee to achieving an impetus for a broader educational outlook, they were soon to be disappointed. The committee was primarily preoccupied with underscoring the importance of a solid training in basic engineering and sciences. Owe himself was quite clear on his view of the engineering education. In his opinion, the NTH should "educate engineers, not economists".[26] Thus, the committee would only recommend a 120-hour course in business administration. Most of the students' time and effort had to be directed towards the ever-expanding field of science and technology. The conclusions of the Owe Committee became a road map for reform at the NTH in the 1950s. As Matthew Wisnioski observed in the case of American engineering education before the 1960s, the majority of the engineering educators in Trondheim retained the view that "the non-technical was non-critical".[27] Hence, this first wave of post-war educational reform indicated a departure from the German *Hochschule* tradition (mixed with homespun polytechnic ideals) towards a more streamlined and specialized engineering education, inspired, to some extent, by American institutions.[28]

---

24 Hanisch and Lange, *Vitenskap for industrien* (1985): 221.

25 Georg Brochmann (ed.), *Vi fra NTH. De første 10 kull* [*Us Who Went to the NTH. The first 10 Classes*] (Stavanger: Dreyer 1934): 141.

26 Hanisch and Lange, *Vitenskap for industrien* (1985): 221.

27 Matthew H. Wisnioski, "'Liberal Education Has Failed': Reading Like an Engineer in 1960s America", *Technology and Culture* 50 (2009), no. 4: 753–782.

28 An alternative view of this shift towards American ideals is presented in a recent, unpublished study: Pål Nygaard, *Profesjonalisering mellom Bildung og Engineering. En studie av de norske ingeniørenes profesjonshistorie 1930–1970* [*Professionalisation between* Bildung *and Engineering. A Study of the Professional History of Norwegian Engineers, 1930–1970*] (Oslo: Kolofon Forlag / Oslo & Akershus University College 2012).

The Academic Drift of the NLHT

The Trondheim teachers' college (NLHT) was caught *in limbo* after the Second World War, while the development of the Norwegian school system was being revised. In the 1930s, the NLHT had been threatened by competition from a number of institutions that offered continuing education for schoolteachers. Just after the war, in 1945, the rector of the NLHT, history professor Arne Bergsgård, argued for the relocation of the institution to Oslo, as the basis for a national institute of pedagogy (see figure 6.3).[29] The idea to transform the small-scale school into an ambitious national institute was perhaps a surprising move since Bergsgård had thus far not received any support to strengthen the NLHT in Trondheim, neither from the Ministry of Education, nor from the government.

FIGURE 6.3 *Professor in history, Arne Bergsgård, was a main strategist for the development of the teachers' college in Trondheim from the early 1930s to the early 1950s. He was also much appreciated as a public speaker. The photo shows Bergsgård in the radio studio of the Norwegian National Broadcasting, 1939.*
PICTURE COURTESY OF THE HILFLING-RASMUSSEN ARCHIVE NO. 3625701, NTNU UNIVERSITY LIBRARY.

29 Arne Bergsgård, "Noregs lærarhøgskule [The Norwegian Teachers' College]", *Norsk pedagogisk tidsskrift* 29 (1945): 193–222.

The move to Oslo did not take place, and the small staff at the farmhouse at Lade, just outside of Trondheim, continued to give their courses as best they could. From 1950, a summer course in English was provided that quickly became very popular. Gradually and grudgingly, Bergsgård changed his strategy. Indeed, the summer course helped to build up local support for the NLHT in Trondheim, but Bergsgård worried that the college would be reduced to a mere summer school. Soon after, the local parliament representatives addressed plans for an expansion of the NLHT. They demanded that the make-shift conditions at the college be brought to an end, and that instead the institution was to be included in the future planning of the Norwegian educational system. In general these demands were met with the restructuring of the public school system in the late 1950s, which paved the way for a longer and more advanced teacher training.

In the mid-1950s, a window of opportunity was opened for expanding the activities at the NLHT. Again, a major restraint was the lack of building facilities, and to remedy this, a brand new campus was inaugurated at Rosenborg in Trondheim in 1960. The strengthening of the teacher education brought results. Overnight, the NLHT grew from 60 to almost 400 students, so the Rosenborg campus was cramped from the outset. A last-minute change in the plans for an expanded institution had led to the inclusion of facilities for teaching chemistry, biology and physics. This was part of the NLHT's strategy to become a fully-fledged institution of higher education. However, the department for the education of science teachers soon developed into a source of contention. Until then, the courses in mathematics and natural sciences for teachers at NLHT had been provided by the NTH professors and by biologists working at the Science Museum owned by The Royal Society of Sciences and Letters. In 1963, the NLHT received their first own professors in physics, mathematics and chemistry. Formally these professors were still employed at the NTH, but they soon began to act independently from the NTH, in favour of the NLHT policy towards a liberal university education.

The University Idea in Post-War Trondheim

During the post-war years, ideas about a university in Trondheim gradually emerged. In 1950, the president of the Royal Society, NTH professor Thorolv Vogt, had envisioned what he called a "Universitas Nidarosiensis". Vogt's use of the medieval name for Trondheim, Nidaros, indicated a classical university ideal. He was aware of the possibilities for building on the existing traditions within science, technology and the humanities in Trondheim, but also took into account the few individual "learned representatives of other faculties such as theology, medicine and agriculture".[30] This reference to these faculties was included in his speech probably more as a friendly gesture towards some specific members of the Royal Society, rather than as a realistic proposal. Still, the idea to establish a faculty of medicine in Trondheim eventually started to take hold. Several influential actors followed Vogt's university idea, although not always along his classical terms. The rector of NLHT in the 1950s and 1960s, Olav Næs, proposed the idea of a university in Trondheim on several occasions.[31] In 1959, the Minister of Ecclesiastics and Education, Birger Bergersen, referred to a university in Trondheim in one of his speeches and in the same year the chairman of the parliament committee for education stated: "We are all taking into account the new university."[32] Thus, the intention of establishing a university in Trondheim had become a political reality.

Less clear were the motives behind the plans for this university. The city of Bergen had obtained a university in 1946, building on the scientific traditions at the Bergen Museum.[33] The long-standing rivalry with Bergen may have been a local motive for wanting a university in Trondheim as well, though it was more likely that the university dream was motivated by an "academic drift" among some of the leading scholars and scientists in Trondheim. A university would mean more research and more emphasis on graduate education. For members of the central government, a university would be a response to the political agenda of growth within the sector. Even if Norway was a relatively slow starter in the post-war scaling up of research and higher education, the international trend towards mass education clearly influenced national policy.

In 1960, the government appointed a committee that would make an assessment of the needs for expanding higher education in Norway. Chaired by the economist and labour party politician Per Kleppe, the committee paved the way for a large-scale expansion of higher education in Norway. In its report from 1961, the committee supported the plans for a university in Trondheim starting from the existing institutions, viz. the NTH, the NLHT, the Royal

30 Quoted in Hans Midbøe, *Det Kongelige Norske Videnskabers Selskabs historie 1760–1960, 2. bind* [*The History of the Royal Norwegian Society of Sciences and Letters 1760–1960. Vol. 2*] (Trondheim: DKNVS 1960): 237.

31 Kirkhusmo, *Akademi og Seminar* (1983): 112.

32 Quoted in Kirkhusmo, *Akademi og Seminar* (1983): 106.

33 For a history of the University of Bergen, see Astrid Forland and Anders Haaland, *Universitetet i Bergens historie, bind 1.* [*The History of the University of Bergen. Vol. 1*] (Bergen: Universitetet 1996).

Society's Science Museum and the university library. However, the Kleppe Committee did not suggest that a university should replace these existing academic institutions. Instead, they believed that a university in Trondheim ought to be different from the ones in Oslo and Bergen. "There is no future premise for the new university that it should cover all of the most important fields of science in the same manner as our two existing universities."[34] Thus, the Kleppe Committee report basically recommended maintaining the existing schools under the university label.

This view of the future university corresponded with ongoing discussions about the organization of higher education in Trondheim. Most of the professors at the technical college were reluctant, sometimes even hostile, to the prospect of joining a university. Their goal was to consolidate the NTH as the nation's most important provider of education and research within the engineering disciplines and architecture, in addition to strengthening the collaboration with Norwegian industry. When asked about the university's prospects, the rector of NTH, Sigurd Per Andersen, stated that the technical college would not mind a "rational collaboration" between Trondheim's scientific institutions. In his opinion, the main aim was "to appropriate what already is in existence". In other words, Andersen and his fellow professors wanted to preserve and consolidate their technical college, yet leave open the possibility of a closer collaboration with other institutions.

In the same year of 1960, the university discussion was broached on the occasion of the 50th anniversary celebration of the NTH. The author of the jubilee volume was Olaf Devik, a former lecturer in physics at the NTH and the NLHT and now a senior counsellor at the Ministry of Education. This resulted in Devik promoting the interests of both the NTH and the ministry. In the conclusion of his book, he envisioned a situation in Trondheim that bore resemblance to "the Anglo-Saxon pattern of autonomous colleges forming a university". Such a structure might offer good opportunities for collaboration between the technical college and the emerging humanities at the teachers' college. So, the wish among engineers and architects could be met to study "human relations", such as the psychological and pedagogical challenges of work-related cooperation and well-being.[35] Devik's use of the concept of

34 *Universitets- og høgskolekomiteen av 1960 ('Kleppe-komiteen'). Innstilling om den videre utbygging av våre universiteter og høgskoler* [*The 'Kleppe Committee' Report on the Development of Higher Education in Norway*] (Oslo: Det Mallingske boktrykkeri 1960): 84.

35 Olaf Devik, *N.T.H femti år. Norges tekniske høyskoles virksomhet 1910–1960* [*Fifty Years of NTH. The Activities of the Norwegian Technical College 1910–1960*] (Oslo: Teknisk Ukeblad 1960): 322.

"human relations" was partly an attempt to bridge the gap between engineering and the humanities, but it also illustrates how non-technical subjects within the social sciences, arts and humanities were seen as ancillary to the education of "well-rounded engineers". To a large degree, this way of thinking indicated that the engineering sciences welcomed university collaboration as long as it was on their terms, and with the premise of preserving the technical college within a loosely knit university structure. In consequence, the professors' council at the NTH voted unanimously in favour of establishing a university in 1961 on the terms drawn up by the Kleppe Committee. Their support not least originated from the strong emphasis that the committee had put on the needs for expansion of the NTH.

Yet, some voices of dissent could still be heard. For some considerable time, the maverick professor of architecture, Arne Korsmo, had opposed the confinement of the architectural education within the technical college. As an internationally renowned architect, Korsmo was wholly disatisfied with what he considered to be a largely narrow-minded engineering school. In a letter to his colleague at the Harvard Graduate School of Design, Jose L. Sert, he complained about being confined within a building full of engineers, and compared it with working "in a factory".[36] His idea was to establish a separate centre for architecture within an umbrella university structure, which would direct itself towards a large part of Northern Norway. Even if Korsmo came quite close to realize his separatist vision, he was quite isolated in his attacks on the NTH.[37] A large majority of the professors remained true to the unity and autonomy of the technical college.

University Integration from above

During the first half of the 1960s, leading representatives from the various academic institutions in Trondheim negotiated over the details of the proposed university. True to form, in 1963 another committee was given the task of finding solutions for the organization of a university in Trondheim. This time

36 Arne Korsmo, "Letter to Jose L. Sert (1956)", quoted in Nina Berre, *Fysiske idealer i norsk arkitektutdanning, slik den ble praktisert ved Arkitektavdelingen, Norges tekniske høgskole i perioden 1945–1970* [*Physical Ideals in Norwegian Architectural Education as Practiced at the Faculty of Architecture, Norwegian Technical College between 1945 and 1970*] (Trondheim: PhD Dissertation NTNU 2002): 178.

37 Norwegian National Archives: RA/S-5478/D/L0001: Arne Korsmo, *Letter to Leif Wilhelmsen, director general at the Ministry of Education* (07/10/1963).

the committee was chaired by Leif Wilhelmsen, director general at the Ministry of Education, but a majority of the members represented the local academic institutions. The primary challenge was how to deal with the fact that there existed three academic institutions dealing with natural sciences in Trondheim. Physics, chemistry and mathematics were taught both at the NTH and at the NLHT. And moreover, these two institutions were also engaged in biology, a field that so far had been one of the scientific pillars of the Royal Society's Science Museum (the other being archaeology). The representatives of the central government within the committee immediately saw the potential for increased collaboration in teaching science courses, and so saving resources. Especially Wilhelmsen himself made a strong appeal for a common science curriculum in Trondheim. However, he realized that for the time being this was really not an option, mainly because the engineering education in Trondheim seemed to represent "an extreme form of regulated college education, with a maximum of course load and a minimum of time for individual study".[38] Still, in his mind, the introductory part of the engineering study course consisted to such a large extent of general science that, over time, it should be possible to find a common ground with the science departments at the teachers' college.

This view was strongly opposed by the NTH faculty members, who agreed that science and mathematics classes for engineering students always would have to be quite different from those given in a university setting. First of all, engineering studies at the NTH lasted only four and a half years, which meant that the course plan had to be more intense than at a six-year university study. Secondly, there was no way of hiding the fact that many of the students at the science departments of NLHT had underachieved at earlier stages of their education, and thus would fail to meet the high entry grade levels of the more prestigious NTH. According to a memorandum by two members of the committee, NTH rector Arne Selberg and Otto Bastiansen from Oslo, only very few arguments were in favour of a common general introduction in science studies for engineers and science majors at the NLHT. Moreover, as far as they knew, the tendency at institutions abroad did indeed go in the direction of more specialization in science education.[39]

With regard to the question of the organization of a university, the committee members agreed on recommending a university, but disagreed on the kind of university that should be established. The representative of the NLHT, the

38 Norwegian National Archives: RA/S-5478/D/L0001: Leif Wilhelmsen, *Draft to memorandum* (08/12/1962).

39 Norwegian National Archives: RA/S-5478/D/L0001: Arne Selberg and Otto Bastiansen, *Draft to memorandum* (08/12/1962).

professor of English Eva Sivertsen, joined a majority in the committee in favour of a loose organization comprising of the existing institutions, with an additional school of medicine that should be established as an entirely new entity.[40] The NLHT position in this discussion oscillated between the drift towards academic freedom promised by a university status and the traditional (more practically oriented) task of educating schoolteachers. Strong signals came from several teachers' organizations against succumbing to the academic forms of education, as they feared that this would imply a theoretical, research-oriented education, far removed from the everyday practice of schoolteachers. This view was also shared by many of the NLHT students, who in the early 1960s were still predominantly schoolteachers seeking continuing education. In the end though, the resistance against becoming (part of) a university was not very pronounced among the NLHT professors and members of the administration.[41]

In general the minority position in such a committee would be of little interest. In the case of the Wilhelmsen Committee however, it did become very important. The minority consisted of chairman Wilhelmsen himself, backed up by Erling Sivertsen, who represented the Science Museum. Together they argued for an integrated university in Trondheim, with a traditional structure including the faculties of humanities, social sciences, natural and engineering sciences and medicine. From their point of view, this was the best way to secure a well-functioning university organization in Trondheim. In their argumentation, Wilhelmsen and Sivertsen evoked the classical term "the idea of a university" as an institution joining all sciences and scholarly subjects of study. If such a universal idea of a university had ever existed, in the beginning of the 1960s it was increasingly challenged by new ideas such as Clark Kerr's "multiversity".[42] While there is nothing in the Trondheim discussion that indicates any knowledge of such new thoughts, it is clear that many of the key figures were looking for alternative ways of applying the term 'university' to the specific local situation.

---

40 *Innstilling om opprettelse av et universitet i Trondheim fra en komité oppnevnt 6. april 1962. Innstillingen avgitt 15. januar 1964* [*Report on the Establishment of a University in Trondheim from a Committee appointed 6 April 1962, submitted 15 January 1964. 'The Wilhelmsen committee'*] (Oslo: Kirke- og undervisningsdepartementet 1964): 81.

41 Kirkhusmo, *Akademi og Seminar* (1983): 113–115.

42 Clark Kerr, *The Uses of the University* (Cambridge, MA: Harvard University Press 1963). With regard to "The Idea of a University", see John Henry Newman, *The Idea of a University* (London: Longmans, Green, and Co. 1907) and Sheldon Rothblatt, *The Modern University and its Discontents: The Fate of Newman's Legacies in Britain and America* (Cambridge: University Press 1997).

The majority recommendation in the Wilhelmsen Committee Report gained much support from the local institutions during the round of hearings, although other institutions in the country were more reluctant towards the proposed solution. The statements from, for instance, the University of Oslo favoured a stronger theoretical emphasis in the engineering education in Trondheim, and that this would shed new light on an integrated science education. In other words, the University of Oslo recommended an integrated model along the lines of the minority in the Wilhelmsen Committee, as this would make collaboration between the universities much easier.[43] This assessment in particular gathered much weight within the Ministry of Education, resulting in the official policy supporting a fully integrated university in Trondheim, instead of a loose assembly of colleges. In a parliament proposition, the ministry "assumed that the reluctance against a larger degree of integration than as proposed by the majority now has abated at the two colleges concerned." Therefore, according to the ministry, it would be feasible to come up with practical solutions for a university with a strong central authority, while simultaneously "preserving valuable traits" of the existing college systems.[44] What this actually implied still remained to be decided, but it is clear that the ministry was presenting a misleading image of the situation in Trondheim to parliament. Indeed, the opposition towards an integrated university nullifying the traditional colleges had by no means abated.[45] Nevertheless, on 28 March 1968 the Norwegian parliament voted for the establishment of an integrated university in Trondheim, but left it up to the local institutions themselves to sort out an organizational structure for this new institution, a task that both the Kleppe and Wilhelmsen Committees had failed to accomplish.

Trondheim in the "Year of Revolution"

Thus far, the students had only sporadically voiced their opinions on the university reform discussions in Trondheim. This was partly due to the fact that they had neither a formal say in the governing bodies of the colleges nor

43 *Om opprettelsen av universitetet i Trondheim* [*On the Establishment of the University in Trondheim*]. *Royal Proposition, St. prp. no. 79, 1966–1967*. (Oslo: Kirke- og undervisningsdepartementet 1967): 7.

44 *Royal Proposition, St. prp. no. 79, 1966–67* (1967): 54.

45 Johannes Moe, *På tidens skanser* [*On the Bulwark of Time. Johannes Moe's Memoirs*] (Trondheim: Tapir 2000): 84.

any representatives in the various committees mentioned so far.[46] According to the 1936 NTH law, which was still in operation in the 1960s, students were to elect representatives on the faculty and college level. It was only in 1948 that the professors' executive council acknowledged the students' right to be represented in the council meetings, and then even only in "matters of special concern for the students".[47] At the end of the 1960s, this was about to change. Students started to protest against top-down reform, while simultaneously demanding reform on their own terms.

Thus far, much of the historiography of '1968' has been emphasized the events at major universities, such as Berlin, Paris, Rome and California, where violent clashes between radical student activists and the police caused headlines in news reports all over the world. Up to today the causes and effects of these students' revolts are still a matter of much debate.[48] Much less attention has been given to the situation at universities outside the central hubs of higher learning that often experienced less violence and confrontation during the 1960s. Nevertheless, many of these places also witnessed a strong urge for reform, mostly combined with a radicalization of the student movements.[49]

In his history of the students' society in Trondheim, Jan Thomas Kobberrød muses: "Was 1968 also a year of revolution in Trondheim's student community? Our answer must be no."[50] Kobberrød may be right in denying that there was ever a student revolt in Trondheim. Compared to the events in Paris in May '68, there is little to tell about Trondheim in terms of revolution or societal upheaval. But then again, not many places can be compared to Paris. While 'Paris May '68' has been steeped in myth, there is little doubt that few sites witnessed a similar uproar by both students and workers. As Kristin Ross has observed, "only in France, and to a certain extent in Italy, did a synchronicity or

---

46 The Owe Committee of 1952, in which two of the six members were students, was an exception in this regard.

47 Central archives of The Norwegian University of Science and Technology (NTNU HA): File 406: *Minutes from the Professors' Executive Council, Professorutvalget sak 106/48.*

48 Paulina Bren, "1968 East and West. Visions of Political Change and Student Protest from across the Iron Curtain", in: Gerd-Rainer Horn and Padraic Kenney (eds.), *Transnational Moments of Change: Europe 1945, 1968, 1989* (Maryland: Rowman & Littlefield Publishers 2004); Ingo Cornils, "Utopian Moments: Memory Culture and Cultural Memory of the German Student Movement", in: Ingo Cornelis and Sarah Waters (eds.), *Memories of 1968: International Perspectives* (Bern: Peter Lang 2010) and Kristin Ross, *May '68 and its Afterlives* (Chicago: University Press 2002).

49 See Egil Førland (ed.), "Special issue on 1968", – special issue of *Scandinavian Journal of History* 33 (2008), no. 4: 317–502.

50 Kobberrød, *Engasjement og Begerklang* (2010): 212.

'meeting' between intellectual refusal of the reigning ideology and worker insurrection occur."[51] The synchronicity between intellectual and worker revolts in French and Italian cities made them the "eyes of the storm". However, according to Jeremi Suri, the 1968 "global wave of urban protests produced a crisis of authority in nearly every society."[52]

The students' involvement in debating higher education in the late 1960s was preceded by a left-wing radicalization that leaned on a transnational pattern of activist ideology, rhetoric and practice. The Cold War armaments race, the brutality of the Vietnam War and the dread of Soviet imperialism had fuelled a young generation with discontent that helped to form a language of dissent.[53] For many young Norwegian intellectuals, China's Cultural Revolution was a fascinating sign of Maoism as a new road ahead.[54] The students' society in Trondheim had a long history as an arena for political struggle, but the ideological climate from the late 1960s onwards became so harsh between the various factions within the society that both moderate students and senior academic members avoided the meetings. In the students' fortnightly newspaper *Under Dusken,* left-wing and right-wing students led irreconcilable debates over a wide range of issues.[55] In 1968 the topic of Norwegian university reform and student rights was widely commented upon by the Trondheim students.[56]

Protest against the Ottosen Committee Reports

In his 1969 pamphlet, *Students and Bureaucracy: The Ottosen Committee Carved and Skewered,* Harald Berntsen, who was a left-wing student at the University of Oslo, summarized the criticism raised by many of his fellow students against the planned reform of higher education in Norway.[57] Students like Berntsen

51 Ross, *May '68 and its Afterlives* (2002): 4.

52 Jeremi Suri, *Power and Protest. Global Revolution and the Rise of Detente* (Cambrige: Harvard University Press 2003): 164.

53 Martin Klimke and Joachim Scharloth (eds.), *1968 in Europe. A History of Protest and Activism, 1956–1977* (New York/Hampshire: Palgrave Macmillan Transnational History Series 2008); Suri, *Power and Protest* (2003): 94.

54 For a history of Norwegian Maoism, see Hans Petter Sjøli, *Mao min Mao. Historien om AKPs vekst og fall [Mao my Mao. The History of the Rise and Fall of the AKP]* (Oslo: Cappelen 2005).

55 Kobberrød, *Engasjement og Begerklang* (2010): 216.

56 E.g. the editorial "Vår egen situasjon [Our Own Situation]", *Under Dusken* (31/08/1968).

57 Harald Berntsen, *Studenter og byråkrati: Ottosen-komiteen partert og anrettet på spidd [Students and Bureaucracy: The Ottosen Committee Carved and Skewered]* (Oslo: Pax forlag 1969).

felt that they had several reasons to be concerned about the educational reforms proposed by the so-called Ottosen Committee[58] that produced a series of reports on higher education from 1966 onwards. For instance, a shorter and more pre-programmed scheme to obtain a degree would mean that students had less time for reflection and extra-curricular studies. Moreover, they would no longer be able to engage in political work. Instead, as Berntsen put it, higher education would serve as an instrument for conserving the established status quo, since the public and private demand for technically skilled employees would be the determining factor of higher education. The Ottosen Committee proposals could no longer be discussed solely in terms of administrative problems concerning how to organize higher education in practice. Rather, the questions arising focussed increasingly on the role of higher education within society.

In his criticism, Berntsen was inspired by Jürgen Habermas' protest against the reform proposals of the West German science council, which had launched similar proposals for rationalization and differentiation in higher education.[59] In the German case, however, students stood against both the professorial bodies and the authorities. They had to position themselves against either the ideology of rationalization or the ivory tower mentality of the aloof professors. In contrast, historian Fredrik W. Thue has found in the case of the University in Oslo that the push for democratic reform came from within the university itself. Only very few Norwegian professors defended their privileges in highbrow philosophical terms.[60] Students in Oslo therefore found allies amongst their teachers in their criticism against a system-conformist kind of reform, such as that proposed by the Ottosen Committee. A strongly regulated form of education, with preset standards for the duration of a degree programme and a division of labour between vocational colleges in the district and the central universities, was seen as contrary to the (undoubtedly mythologized) Humboldtian ideals of academic freedom for students and teachers alike.[61] Before 1968, no one in Norway had really paid any attention to the figure or the

58 The committee was named after its president, Kristian Ottosen.

59 Hans Joachim Hahn, *Education and Society in Germany* (Oxford/New York: Berg publishers 1998): 125–126.

60 Fredrik W. Thue and Kim G. Helsvig, *Universitetet i Oslo 1945–1975. Den store transformasjonen* [*The University of Oslo 1945–1975. The Great Transformation*] (Oslo: Unipub forlag 2011): 333–334.

61 This "division of labour" was introduced on a broad basis in Western Europe during this period through the so-called "binary model" of higher education, see Svein Kyvik, "Structural Changes in Higher Education Systems in Europe", *Higher Education in Europe* 29 (2004), no. 3: 393–409.

ideas of Wilhelm von Humboldt, but through various publications from, among others, the philosopher Hans Skjervheim, the "father of the modern German university" soon became a household name and a metonymy for the ideal type of university. For Skjervheim, the central concept was *Bildung*, what he defined as a free and open intellectual reflection process that defied any bureaucratic regulation or standardization.[62] According to Thue, the situation in Oslo was quite astonishing because it seemingly implied a surpassing of the general contrast between an elitist defence of university tradition on the one hand, and a radical participatory, democratic ideal of higher education on the other hand.[63]

The situation in Trondheim had some common traits with that in Oslo, particularly regarding the cry for reform from within. At the same time, there is little doubt that the development in Trondheim was quite different from that in Oslo, both in terms of its origin and of how the conflict over the university reform unfolded in the 1960s and 1970s. The level of conflict between the students and the various authorities was substantially higher in Oslo than in Trondheim in 1968 and the ensuing years. One of the reasons for this was that the major educational institution in Trondheim, the NTH, was explicitly kept out of the reforms suggested by the Ottosen Committee, at least in its main report.[64] According to this report, the NTH was one among several other higher college institutions in Norway that worked well in terms of students graduating on time and with very high completion rates.[65] In his pamphlet, Berntsen distrusted the Ottosen Committee's lauding appraisal of well-working colleges such as the NTH. The committee had not bothered to examine these institutions in terms of "the level of academic reflection, the degree of cultural self-understanding and the extent of political and other extra-curricular activities". In Berntsen's opinion, the committee would rather advocate a kind of

62 Hans Skjervheim, *Ideologi og universitetsreform* [*Ideology and university reform*] (Oslo: Aschehoug 1969/1996). Also available electronically at http://urn.nb.no/URN:NBN:no-nb_digibok_2010061408006 (date accessed 13/09/2011).

63 Thue and Helsvig, *Universitetet i Oslo 1945–1975* (2011).

64 Only a sub-committee for technical education was appointed, which was chaired by the former NTH rector Andersen and which eventually became a target for criticism by the students in Trondheim. See *Komité for høgre teknisk utdanning. Innstilling om teknisk utdanningsstruktur i Norge i 1970–1980-årene* [*Report on the Structure of Technical Education in Norway in the 1970s and 1980s*] (Oslo: Kirke- og undervisningsdepartementet 1969).

65 *Innstilling nr. 1 fra komitéen til å utrede spørsmål om videreutdanning for artianere og andre med tilsvarende utdanning* [*Report from the Higher Education committee 'The Ottosen Committee,' Report no. 1*] (Oslo: Kirke- og undervisningsdepartementet 1966): 19.

university exclusively appropriated to the industrial society, akin to being "a factory for technical knowledge".[66]

Students and Democratic Struggle in Trondheim

The academic community in Trondheim was by no means insulated from the university discussions in the Norwegian capital, nor from those unfolding elsewhere in the world during the late 1960s and early 1970s. In fact, students in Trondheim turned out to be just as critical towards their own school as Berntsen was in his pamphlet. From 1968 onwards, students at the NTH criticized their institution along three lines, sometimes intersecting with each other, at other times causing conflict among the students themselves. Firstly, they launched a plea for a thorough study plan reform, which basically was a criticism of an existentialist kind directed towards the (supposed) lack of meaning and purpose in the education; secondly, the students were striving for democratic reforms that stemmed from a frustration over the students' lack of formal influence in the governing bodies of the college; thirdly, the NTH was criticized from an ideological perspective, based on the assumption that the institution had become one of the cogs in the imperialist machinery of Western capitalism.

The actual year of 1968 itself turned out to be a rather quiet one in Trondheim in terms of student protest on the streets, campus occupations, teach-ins and similar modes of action that were widespread in other places. Still, a few episodes did mark the link with these global events. In the early spring of 1968 for instance, the German student leader Rudi Dutschke visited Trondheim on a short Norwegian tour. The leading conservative daily, *Adresseavisen*, were very critical about his visit, even going so far as to accuse Dutschke of being a criminal. After the attempted assassination of Dutschke in April 1968, the Trondheim students organized a protest march against local news coverage, accusing *Adresseavisen* of being an accomplice to the detested German *Springer Press* (see figure 6.4).[67]

Another episode that resembled the transnational pattern of 1968 student action was a protest of NTH students in technical physics at the end of the spring term. The students refused to take a course in machine parts because they considered it an obsolete remnant of a polytechnic tradition now being swept away. Because two of the leading students were members of the local

66 Berntsen, *Studenter og byråkrati* (1969): 14.

67 "Røde Rudi [Red Rudi]", *Adresseavisen* (13/03/1968). For the coverage of this episode, see *Adresseavisen* (13, 17, 19, 20, 22/04/1968).

FIGURE 6.4 *1968 in Trondheim: students protesting outside the offices of local newspaper* Adresseavisen *after the attempted assassination of the German student leader Rudi Dutschke. The banner in the middle of the photo reads: "Why does Adresseavisen lie about Dutschke?"*
PICTURE COURTESY OF *FOTOGJENGEN, STUDENTERSAMFUNDET I TRONDHEIM*, FOTOGJENGEN.SAMFUNDET.NO (DATE ACCESSED 19/06/2014)

Socialist Youth Association (SUF), an overzealous *Under Dusken* journalist jumped to the conclusion that "the student revolts in West Germany [and] France obviously had brought to attention the question of student democracy at the NTH." However, the two SUF members themselves were reluctant, even hostile, to the assumption that this had anything to do with international events. They were eager not to bring their relationship with their professors to a head; a relationship that, according to the students, was much better than at German institutions. As long as the professors and the administration acted "rationally", there were few reasons to fear any revolts in Trondheim.[68]

The episode nevertheless illustrates that the students intended to act against the lack of reform of both study plans and university governance. During 1968, an increasing number of students voiced their criticism over the engineering and architecture courses. One of them, Kjell Rabben, spoke on behalf of many of his fellow students when giving a critical review of Rector Arne Selberg's inaugural speech for the autumn term of 1968. Rabben claimed that even if a

68 Unknown author, "Hvorledes fungerer studentdemokratiet på NTH? [How does student democracy work at the NTH?]", *Under Dusken* (11/05/1968): 158–160.

society establishes a technical college, this does not give the society any right to "manipulate its students". Furthermore, Selberg was described as a representative of the "static society", whereas Rabben's own vision was clearly different:

> We want a critical university able to create an impulse for societal change. In order to achieve this, we need to liberate ourselves from the manipulation. To do so we need power, student power. This is really only a matter of democratization. We want power to decide on ourselves, on our working situation and our education.[69]

Rabben's critical remarks demonstrated that among the students at the NTH there existed a strongly felt and multi-layered discontent with the educational system. He wanted a university education with time for critical reflection, not a technical school marred by intense cramming. He wanted a meaningful influence of the students, not an uncontested professorial oligarchy; ultimately, Rabben and his fellows urged for a different society.

Whether or not Rector Selberg took these harsh remarks to heart remains unknown. It was clear however that a thorough assessment of the governing structure at the NTH was long overdue. Pressure for reform was growing, and it was not only coming from disgruntled students. Junior staff members were exceedingly demanding more influence in running the college business. By 1968, they had become a force to be reckoned with. While in 1936 there had been only two junior faculty members per professor, the ratio had now soared to more than five.[70] The strains on the existing rules and regulations were evident, and with the new university structure looming, Rector Selberg appointed a committee with the task to put forward suggestions for future governing bodies at the NTH, along with ideas about how these could function within the new university.

This committee notably differed from the previous committees by the fact that it consisted of only three members (plus a secretary): Professor Andreas Tonning as chairman, the lecturer Tor L. Bakken, and Kaare Granheim, as the representative of the students. This meant that for once professorial dominance had been abandoned, even if Tonning probably wielded more real and symbolic power than his fellow members together. The Tonning Committee

---

69 Kjell Rabben, "Teknokraten. Kritikk av en immatrikuleringstale [The Technocrat. A critique of an Inaugural Speech]", *Under Dusken* (14/09/1968): 176.

70 NTNU HA: File 05, folder 2: *Komité oppnevnt av Rektor. Forslag til nye styringsorganer ved NTH [Report on Future Governing Bodies for the NTH 'The Tonning Report']* (12/1968): 5.

became influential on several levels, as it worked swiftly enough to prevent the kettle of discontent from boiling over. It started working on 30 October, and exactly two months later a report had already been produced. Admittedly, the report was made confidential to those other than the rector and his executive council. However, this was a misjudgement since the report was leaked and so the discussions were given a false start.[71] Some of the recommendations in the report were controversial, but all in all there was a general consensus about the overall direction suggested by Tonning and his two accomplices. Instead of the traditional model of faculties divided into institutes for each professorial chair, the committee recommended a faculty structure based on larger "sections" or departments containing several related disciplines, with their chairs and their laboratories.

Few of the professors protested against the abolishment of the outdated model with its personal institutes, which two decades earlier had led to the witty remark that described the NTH as "a confederation of 40 autonomous republics".[72] More concerns were uttered about the size of the various boards and councils planned by the committee, and whether the potential for interdisciplinary collaboration was catered for well enough. In short, they were actually only discussing details.[73] Potentially controversial recommendations, such as including students' representatives on all levels and allowing the possibility for non-professors to be heads of departments, passed almost unnoticed. Apparently without a fight, the thus far dominant professors gave up their age-old omnipotence. Most of them held a pragmatic view of being in charge of a department, a laboratory or any such institutional body. As long as someone competent was running the business, they would go along with it. Some of the professors probably also felt relieved by the prospect of others stepping in at the helm of things. With regard to the admission of students into the boardrooms, the professors had some minor objections. As long as the student body only had a few representatives, guaranteeing them the support of their peers, it would be just as beneficial to have more students in the governing bodies.

Tonning's most ardent critics came from within the student body itself. Knut Aarvak, a left-leaning former editor of *Under Dusken*, lashed out against

71 NTNU HA: File 05, folder 2: *Minutes from an extended Professors' Executive Council meeting* (07/02/1969).

72 Stig Kvaal, "Etableringen av SI og SINTEF [The Establishment of SI and SINTEF]", in: Jon Gulowsen, *Bro mellom vitenskap og teknologi. SINTEF 1950–2000* (Trondheim: Tapir 2000): 407.

73 NTNU HA: Fome 05, folder 2: *Minutes from a Professors' Executive Council meeting* (17/01/1969).

the committee, charging its members of being "faithful servants of the system". He particularly regretted the committee's rejection of any kind of ideological discussion. Such an explicit dismissal of putting into question the ideological underpinnings of the college system could not be seen as anything else but "monstrous" since this had been the main concern of the students. According to Aarvak, the immediate cause for Rector Selberg to appoint the committee had been the students' addressing of the democratic shortcomings and ideological weaknesses of the college system. But even despite these flaws, Aarvak had to admit that the committee had come up with some substantial improvements in order to create a more democratic college structure.[74]

In all likelihood, Aarvak was wrong in claiming that the students' protest was the main reason for the appointment of the Tonning Committee. Several forces pushed for reform, so both among the professors and the national authorities the development was highly welcome. Indeed, the students would have wanted even more influence if they could have gotten it. They appointed their own committee for democratization, working in parallel with Tonning's. Their report contained a scorching critique of the education at the NTH. An outdated organizational structure marred by pigeon-holing, paternalism, a too heavy workload for the students and too little freedom of choice in the course plans, were only some of the criticism.[75] The Tonning Committee were unable to suggest improvements on all these matters, though eventually the overarching conclusions in their report became part of the NTH's policy, thereby creating room for more democratic and flexible forms to organize the college.

A University Left High and Dry

While the internal differences at the NTH were brought forward in a constructive manner, the negotiations over the university model in Trondheim gradually stagnated. After the Norwegian parliament had voted for an integrated university in Trondheim, the Ministry of Education passed the task of actually creating a good working university to the local institutions themselves. An interim board of directors was set up for the university, which was assigned the task of fleshing out a university based on the existing faculties at the NTH and the NLHT, in addition to the museum, the libraries and a new faculty of

74 Knut Aarvak, "Systemets trofaste tjenere [The Faithful Servants of the System]", *Under Dusken* (29/08/1969): 169.

75 NTNU HA: File 463, folder 1: *Innstilling fra studentenes demokratiseringskomité* [*Report from the Students' Democratization Committee*] (22/11/1968).

medicine. The interim chair of the board would alternate between the two rectors of the two colleges.

At the start, there was an atmosphere of curbed optimism in the conference room. The two rectors, the historian Edvard Bull from the NLTH and Selberg from the NTH were on good terms, and more importantly, they represented their respective college communities with integrity. In 1969, however, new rectors were elected. The new rector at the NLHT was Haakon Olsen, who had graduated as a physicist from the NTH, but who, according to one of the NTH's leading men, had turned out to be almost hostile towards the engineering college.[76] If Olsen was acting against the interests of the NTH, it might be explained by the enormous growth that the NLHT was experiencing during this period and which put a strain on the faculty members and the facilities.[77] Olsen found a good ally in Gunnar Bøe, the new rector at the NTH. Bøe was a professor in economy, and thus not a representative of the hard-core engineering culture. Besides, as a former member of the Norwegian government and a prominent Labour party politician, he should have been well versed in negotiations such as these. Bøe had been elected as rector in large part due to the fact that as a social scientist he would have a better chance of building bridges across the cultural division between the colleges.[78]

However, this strategy backfired as Bøe and Olsen turned out to agree on several points of great concern for the NTH. It is impossible to deny the fact that most of the professors at the NTH really did not welcome a university. Their strategy was to delay and impede the process as best as they could, making only small symbolic concessions. But when Bøe and Olsen agreed upon a model for a joint university budget in which the university board would control even the funding of the NTH, a riot almost broke out among the engineering professors. One of them was Johannes Moe, who chaired the NTH's central study plan committee. Accompanied by Egil Abrahamsen, the general director of *Det norske Veritas* and chair of the NTH industrial cooperation committee, he travelled clandestinely to Oslo for a meeting with the parliamentary committee in charge of higher education. With Abrahamsen at his side as a powerful symbol of the symbiotic relationship between engineering education and industry in Trondheim, Moe convinced the politicians about the necessity of preserving the NTH's autonomy for the time being. Only when the aims and organization had been agreed upon could the NTH relinquish its power over their education to a university. According to Moe, Bøe was infuriated when he

76 Moe, *På tidens skanser* (2000): 85.

77 Kirkhusmo, *Akademi og Seminar* (1983): 164.

78 Moe, *På tidens skanser* (2000): 85.

discovered the subversive action, and Moe was given a thorough dressing down in front of his colleagues. It was probably the best promotion imaginable for Moe's candidacy for the upcoming election for the NTH rectorship.[79]

The episode of Moe campaigning for his engineering college by using all possible means illustrates clearly how difficult it would be to find a solution for the establishment of a university in Trondheim. With Moe as the NTH rector from 1972 to 1975, the negotiations were kept on civil terms but in the end neither he nor the NTH community as a whole would budge an inch. A separate structure committee looking at organizational models had been put to work from 1969. Chaired by the venerable director of the Students' Welfare Organization (*Studentsamskipnaden*), NTH alumnus and war-hero, Finn Meland, this committee should have had the best conditions for coming up with an overall, acceptable suggestion. The structure committee did indeed bring to the fore a multitude of assessments and reports on university models. British "plate-glass" universities worked as an inspiration for physical structuring, while some of the students involved in the process came up with a structure based on ideas taken from, for instance, integral algebraic equations.[80] But, alas, these ideas came to naught. In the late spring of 1973, Meland was informed that the NTH students' representatives in the committee, who he had given time off from committee work to study, had in fact been hiding at a local hotel with other NTH staff members discussing the future of the NTH. Meland was highly disappointed, and insisted that the committee members refrain from this kind of factionalism, as it was not "fair play".[81]

Was it naïve to expect fair play in these negotiations? Probably yes. Too much was at stake for the NTH to relinquish anything of substance to a university. But at the same time, one might consider the dominant NTH's reluctance towards university integration as somewhat rigid, lacking imagination and possibly testifying to a lack of self-confidence on behalf of the academic resilience of the college's education. While the discussions in the closed boardrooms crawled to a halt, both of the colleges grew in size. In particular, the NLHT witnessed a remarkable expansion from 1.140 students in 1968 to 3.541 in

79 Moe, *På tidens skanser* (2000): 84–88.

80 Michael Beloff, *The plateglass universities* (London: Secker & Warburg 1968); Kjell Askeland and Per Ivar Maudal, *"Universitetet": fragmentarisk innføring i planleggingens problematikk* [*"The University": A Fragmentary Introduction to the Problems of Planning*] (Trondheim: Published by the authors 1969).

81 NTNU HA: Archival no. 022.2.5, file "Strukturkomiteen," folder "Interne notater": Finn Meland, *Letter to the members of the Structural Committee and the Chairman of the Interim Board* (02/06/1973).

1979. During the same period, the number of staff increased from 144 to 331.[82] In comparison, the NTH had 2.815 students in 1968 and 3.552 in 1979. This meant that the NLHT had amassed almost the same size as its once so intimidating bigger cousin. The growth mostly took place within the humanities and the social sciences, disciplines that were easier to scale up than the professional education of engineers and architects that demanded more resources. It would also be reasonable to say that the NTH enjoyed a higher status in Norwegian political discourse than the NLHT. These two factors explain why the NTH had a staff of already 1.133 members in 1970 (and which increased to 1.332 a decade later). Additionally, the NLHT had to wait 10 years for a properly sized campus, while the NTH enjoyed a boom related to the budding oil industry build-up.

All in all, while the two competing institutions grew closer in size, the difference in scope, resources and prestige seemed to be widening. The shadowboxing between the institutions in Trondheim went on for several more years, and was only interrupted by some makeshift arrangements to please the impatient but impassive central government. A university was formally established in 1976, although the NTH remained an autonomous entity. It was only in 1995 that a fully integrated university was agreed upon by the politicians in parliament, with the new construction 'The Norwegian University of Science and Technology' operating from 1996 onwards. The old college names were put to rest in favour of the novelty. It is still however too early to conclude that a new united university identity has superseded the old ones.

University History and the Politics of Identity

Many social identities can be found overlapping in an organization like the NTNU. The dialectics of identity construction between individual self-definition and collective community building will normally involve references to history in one way or another. As the anthropologist Jonathan Friedman has observed, social identity as a process does not occur in a vacuum. "As such it invariably fragments the larger identity space of which its subjects were previously a part."[83] At a university, students are socialized into a set of social identities. Yet, as students in Trondheim are faced with this set of possible identities, it is clear that there is a power struggle going on between different academic

82 Kirkhusmo, *Akademi og Seminar* (1983): 142 and 159.

83 Jonathan Friedman, "The Past in the Future: History and the Politics of Identity", *American Anthropologist* 94 (1992), no. 4: 837–859.

sub-cultures over what image of the university students should be presented with. What part of the university's past should be considered most important may be considered a form of "hidden curriculum".[84] In Trondheim this often has to do with the legitimacy of the supremacy of natural science and technology over other fields of study.

When the NTNU was established, the Norwegian parliament gave concessions to the engineering tradition through the adoption of a main profile for the new university. Natural science and technology were to be the pillars of future academic activity in Trondheim. This generated vocal protests from those parts of the university community that felt excluded by this main profile. How could scholars working within disciplines like philosophy, literature and music find room for their work under the dominion of science and technology? Such tensions were intensified by the situation with two main campuses, one called Gløshaugen, the traditional home of the engineering and technology faculties, and another called Dragvoll, housing the humanities and social sciences. This physical separation became a sort of embodiment of the traditional academic gulf between "two cultures". Attempts at reconciling this division by planning to co-locate the whole university in one campus have failed due to a combination of lack of funding and lack of interest from the national government. Even if the situation today is marked by a lesser reciprocal animosity between Dragvoll and Gløshaugen, there is still a long way to go before a shared NTNU identity can be found.

The NTNU jubilee in 2010 had several objectives. One of the problems that the NTNU is facing is to get its name known in the academic and scientific world. The cry for excellence has come through, here as everywhere else. The jubilee was thus used as an occasion to promote NTNU's strengths within research and higher education. But it was also an objective for the jubilee to consolidate and build a common identity internally.

What part did history play in these objectives? In the case of the NTNU it is not obvious that knowledge about the past would be of benefit. Being reminded of former differences of opinion may be seen as opening old wounds. This could be one of the reasons for the belated planning of the jubilee in general and in particular the hesitant decision to produce a work of history on the occasion of the jubilee. The history project was initiated "from below", as a result of the efforts from 2006 on by an almost self-organized group of people

84 Elizabeth Vallance, "Hiding the Hidden Curriculum: An Interpretation of the Language of Justification in Nineteenth-Century Educational Reform", *Curriculum Theory Network* 4 (1973–1974), no. 1: 5–21.

interested in creating a celebration of the various knowledge institutions in Trondheim.[85]

The main question was how to find a satisfactory way to frame a history of a set of such quite different academic institutions, and, as a corollary to that question, satisfactory for whom? Quite early on in the project, the vice-president of the Royal Society, himself a professor from the former NTH, suggested a working title for the history volume, inspired by a quote from Knut Hamsun, a Norwegian Nobel laureate in literature: "A hundred years, but not forgotten – The history of the NTH". Alternatively he suggested we should use the name "Gløshaugen" in the title, as this was familiar to most Norwegian readers.[86] This indicates how difficult it was from the start to communicate that the history of the NTNU was more than 100 years of engineering education, and that Gløshaugen was not the only site of this history. The vice-president's suggestions were probably based on an innocent misunderstanding of the institution's past. It could however also be interpreted as an example of the politics of identity at work: the Royal society was very interested in presenting itself as the sole proprietor of a 250-years old tradition. The fact that the NTNU also contains a museum and a library dating back to 1760 was downplayed in importance.

Writing history cannot be removed from having a political potential. Our contribution is in part to show a common history that in many ways debunks the mythical conception of two cultures at the NTNU, and instead to demonstrate a series of academic, scientific and scholarly practices evolving through this history. Admittedly, our attempt to create an overarching historical framework may run the risk of establishing yet other myths of historical continuity.[87] It would have been considerably easier to write a history solely about the engineering education in Trondheim. It would have meant a shorter time period and a more homogeneous institutional framework.

We have also chosen to discuss more than just the institutional integration processes. Our main contribution will be to present these different institutions' histories alongside histories of academic practices from various fields of knowledge, and to put these stories into a broader context. What are the different *Bildung* traditions that have been active in Trondheim? How has the role of

85 Samarbeidskomité for jubileumsarrangementer i 2010, *Innstilling* [*Recommendations from the 2010 jubilee co-operation committee*] (Trondheim 27/08/2006).

86 Kristian Fossheim, e-mail correspondance with the author (23/11/2008).

87 For a critical assessment of "constructed or evasive narratives" in university commemorative practices, see Picard, "Recovering the History of the French University" (2013).

the engineering professor changed from polytechnic consultant and teacher to specialized researcher with industrial funding? How did scientists and students engage in issues of global awareness like poverty alleviation through education in East Africa, environmental protection and the battling of climate change through knowledge and technology? How did international trends in architecture and urban planning influence the development of buildings and campuses at the university? These and other questions about the world outside were raised in the NTNU history, alongside discussions of the development of an integrated university. Our hope is that discussing a combination of internal and external developments will give the interested public, mainly alumni, students and employees at the NTNU, some pegs on which they can hang their historically based identities.

CHAPTER 7

University History Research at the University of Leipzig

Jonas Flöter

In 2009, the University of Leipzig looked back on 600 years of academic history. The Alma Mater Lipsiensis is the second oldest university in Germany with the longest continuous period of teaching and research. Famous figures such as Gottfried Wilhelm Leibniz, Johann Wolfgang Goethe, Wilhelm Wundt, Wilhelm Ostwald and Werner Heisenberg studied or taught at the University of Leipzig. The 600-year-old history of the university was affected by diverse intellectual, religious, political and social influences. At the same time miscellaneous innovations in the humanities and sciences have been achieved at the university.

At the beginning of the 1990s, the plan to write a new history of the University of Leipzig developed. In 2000, the so-called Senatskommission zur Erforschung der Leipziger Universitäts- und Wissenschaftsgeschichte *was founded. Although the main focus of the article is on the challenges and questions that the commission had to face when writing a new history, in the first section the status of the University of Leipzig within the setting of other Central European universities throughout its history is developed in a little more detail, by way of introduction. The second section deals with the establishment and the work of the commission. Its main result was a five-volume history of the University of Leipzig, which is presented in the third and last section. The development and structure of the volumes is discussed, as well as some of the methodological and scientific problems the commission encountered during the writing process.*[1]

There are many universities in Central Europe that have long-standing traditions. However, only a few of them have a history dating back to the Middle Ages. In 2009, the University of Leipzig celebrated its 600th anniversary. On 2 December 1409, the university was founded by the refectory of the Augustiner-Chorherren-Stifts St. Thomas, the first vice-chancellor being Johann Otto von Münsterberg. However, the foundation of the university was in the first place the initiative of the margrave Frederick IV and his brother William II.[2]

---

1 I owe thanks to Antje Keilholz, Kirsten Parker and Manfred Rudersdorf for all their hints and support with the elaboration of the manuscript.

2 Jacques Verger, "Grundlagen", in: Walter Rüegg (ed.), *Geschichte der Universität in Europa. Bd. 1: Mittelalter* (München: Beck 1993): 49–80 and Enno Bünz, "Gründung und Entfaltung.

 | DOI 10.1163/9789004265073_008

The University of Leipzig soon reached a leading position amongst the universities of central Europe. In the following centuries, it was able to maintain its position and expand. Throughout many societal and political changes, including the dictatorships of the twentieth century, the university managed to maintain its scientific reputation. Today, it is one of the oldest still existing universities in Germany.

The University of Leipzig was often directly affected by national and local crisis situations and faced many challenges and reforms over the years. During the Reformation, the university changed from being an educational institution of the Middle Ages to a university for spiritual, juristic and pedagogical studies to the service of Albertinist Saxony. This was laid down in the university regulations of 1580. Around 1800, the struggle for another university reform started. Notable figures such as the philologist Gottfried Hermann and the philosopher Wilhelm Traugott Krug laid down several plans for it. The university reform that was realised in 1830 replaced the subdivision of the university in four nations (Meissen, Saxony, Bavaria and Poland) with a new, modern faculty structure. The reorganization of the university was accompanied with social reforms within the state of Saxony. The reforms within the university laid down the foundation for a significant development of it during the nineteenth century. Around 1900, the University of Leipzig, along with Berlin and Munich, belonged to the most prominent universities in Germany.

In the first section of this article, by way of introduction, the status of the University of Leipzig within the setting of other Central European universities throughout its history is developed in a little more detail. However, the main focus of the article is clearly on the challenges and questions that the commission for the study of the history of the University of Leipzig (*Senatskommission zur Erforschung der Leipziger Universitäts- und Wissenschaftsgeschichte*) had to face when it was decided to write a new history of the university on the occasion of its 600th anniversary. The second section deals with the establishment and the work of this commission. Its main result was a five-volume history of the University of Leipzig, which is presented in the third and last section. The development and structure of the volumes is discussed, as well as some of the methodological and scientific problems the commission encountered during the writing process.

Die spätmittelalterliche Universität Leipzig 1409–1539", in: Enno Bünz, Manfred Rudersdorf and Detlef Döring, *Geschichte der Universität Leipzig 1409–2009. Bd. 1: Spätes Mittelalter und Frühe Neuzeit 1409–1830/31* (Leipzig: Universitätsverlag 2009): 21–79.

The Status of the University of Leipzig within the Setting of Central European Universities

The foundation of the University of Leipzig was one of the most spectacular events in the history of Central European educational institutions. As a result of fierce hostilities over political and religious questions between the Bohemian King Wenceslas IV and predominantly German professors at the University of Prague, there was a split in 1409 and roughly 2000 professors and students left Prague.[3] Most of them went to the universities of Vienna, Erfurt and Cologne. A small number however accepted the invitation of margrave Frederick IV to found a new university in Leipzig modelled on the University of Prague. The University of Leipzig became in that way the sixth university within the Holy Roman Empire. From the beginning the margrave also supported his state university financially. It was in direct competition with the University of Erfurt, which was founded in 1392 and had already developed into a centre of German humanism.[4]

Only in the age of the Reformation was it possible for the University of Leipzig really to compete against the University of Erfurt. In the same period, more universities were founded: in Wittenberg in 1502 and in Jena in 1547. It was the University of Wittenberg which in particular developed as the heart of the Reformation, due to the presence of the reformers Martin Luther and Philipp Melanchthon. Humanism and the Reformation formed the basis for the gradual promotion of the University of Leipzig. However, at the beginning of the eighteenth century, it faced new competition from the recently founded universities in Halle (1694) and Göttingen (1732/1734). Until the nineteenth century, the University of Leipzig was not at the cutting edge of new developments in humanities and sciences. Nevertheless, the newest scientific developments were usually taken up quickly and attempts were made to convince leading figures in modern scientific areas to move to Leipzig.[5]

---

3 See p. 30.

4 Helmar Junghans, *Der mitteldeutsche Renaissancehumanismus. Nährboden Frühen Neuzeit* (Sitzungsberichte der Sächsischen Akademie der Wissenschaften zu Leipzig, Philologisch-historische Klasse 139, 1) (Leipzig: Hirzel 2004); Günther Wartenberg, "Sachsen und die Universität Leipzig im Spätmittelalter", in: Günther Wartenberg, *Wittenberger Reformation und territoriale Politik. Ausgewählte Aufsätze*, edited by Jonas Flöter and Markus Hein (Leipzig: Evangelische Verlagsanstalt 2003): 241–248.

5 Herbert Helbig, *Die Reformation der Universität Leipzig im 16. Jahrhundert* (Gütersloh: C. Bertelsmann 1953); Günther Wartenberg, "Die kursächsische Landesuniversität bis zur Frühaufklärung, 1540 bis 1680", in: Lothar Rathmann (ed.), *Alma Mater Lipsiensis. Geschichte der Karl-Marx-Universität Leipzig* (Leipzig: Edition 1984): 55–75; Manfred Rudersdorf,

The town of Leipzig itself played an important role in the development of the university. Leipzig belonged to the most important trade centres in Europe and developed into a centre of book printing. This location gave its university an important advantage over its counterparts in Erfurt, Jena, Wittenberg, Halle and Göttingen.[6]

During the nineteenth century, the University of Leipzig reached the climax of its development. Since the reforms of the state of Saxony and the University of Leipzig in 1830 and 1831, the Saxon government undertook intensified efforts in order to raise scientific performance and kudos. This succeeded remarkably well. In the second half of the nineteenth century, countless prominent scientists were brought to Leipzig, such as the classical philologist Gottfried Hermann, the historian Karl Lamprecht, the natural philosopher Gustav Theodor Fechner, the psychologist Wilhelm Wundt, the chemist and later Nobel Prize winner Wilhelm Ostwald (see figure 7.1), the theologian and church historian Konstantin von Tischendorf and Albert Hauck, the jurists Christian Gottlieb Haubold and Carl Georg Wächter, as well as the medical scientists Karl Thiersch and Friedrich Trendelenburg.[7] After

---

"Weichenstellung für die Neuzeit. Die Universität Leipzig zwischen Reformation und Dreißigjährigem Krieg 1539–1648/1660", in: Bünz, Rudersdorf and Döring, *Geschichte der Universität Leipzig 1409–2009, Bd. 1* (2009): 331–519 and Detlef Döring, "Anfänge der modernen Wissenschaften. Die Universität Leipzig vom Zeitalter der Aufklärung bis zur Universitätsreform 1650–1830/31", in: Bünz, Rudersdorf and Döring, *Geschichte der Universität Leipzig 1409–2009, Bd. 1* (2009): 521–574.

6 Henning Steinführer, "Stadt und Universität am Übergang vom Spätmittelalter zur Frühen Neuzeit", in: Detlef Döring (ed.), *Universitätsgeschichte als Landesgeschichte. Die Universität Leipzig in ihren Territorialgeschichtlichen Bezügen* (Leipzig: Evangelische Verlagsanstalt 2007): 25–40. Compare also the conference "Stadt und Universität. Tagung des Leipziger Geschichtsvereins e.V., der Stadt Leipzig, der Universität Leipzig und der Sächsischen Akademie der Wissenschaften zu Leipzig, 16–17. Oktober 2009".

7 These successes are especially emphasized in the institutional university history published on the occasion of the quinquecentennial in 1909, see *Festschrift zur Feier des 500jährigen Bestehens der Universität Leipzig herausgegeben von Rektor und Senat* (Leipzig: Hirzel 1909). It consists of 5 volumes: Otto Kirn, *Die Leipziger Theologische Fakultät in fünf Jahrhunderten*; Emil Friedberg, *Die Leipziger Juristenfakultät, ihre Doktoren und ihr Heim*; *Die Institute der medizinischen Fakultät an der Universität Leipzig*; *Die Institute und Seminare der philosophischen Fakultät an der Universität Leipzig. Die philologische und die philosophisch-historische Sektion*; *Die Institute und Seminare der philosophischen Fakultät an der Universität Leipzig. Die mathematisch-naturwissenschaftliche Sektion.* See also Konrad Krause, *Alma mater Lipsiensis. Geschichte der Universität Leipzig von 1409 bis zur Gegenwart* (Leipzig: Universitätsverlag 2003): 111–242 and Jonas Flöter, *Leipziger Universitätsgeschichte(n). 600 Jahre Alma mater Lipsiensis* (Leipzig: Evangelische Verlagsanstalt 2009): 88–137.

FIGURE 7.1 *Wilhelm Ostwald, Nobel Prize winner for chemistry in 1909.*
PICTURE COURTESY OF WILHELM-OSTWALD-PARK IN GROßBOTEN

FIGURE 7.2
Werner Heisenberg, Nobel Prize winner for physics in 1932.
PICTURE COURTESY OF THE ARCHIVES OF LEIPZIG UNIVERSITY

the First World War, the development of scientific research at the University of Leipzig reached a high point. Firstly, the physicist Werner Heisenberg won the Nobel Prize in physics in 1932 (see figure 7.2) and only four years later his colleague, the chemist Peter Debye, joined him in the select company of Nobel Prize winners (see figure 7.3).[8]

In the twentieth century, under two dictatorial regimes, the university experienced its lowest point of development. During the time of the National Socialist regime, 44 professors and lecturers were fired because of racist or political reasons. The attitude of professors towards the regime ranged from an open acceptance of it, through conforming to the ideologies it held, or to an out and out rebellion towards it. The historian and folklorist Adolf Helbok and the philosopher Arnold Gehlen were two of the main supporters of National Socialism in Leipzig. Conversely, some of their colleagues may not have openly

8 Christian Kleint, Helmut Rechenberg and Gerald Wiemers (eds.), *Werner Heisenberg 1901–1976. Beiträge, Berichte, Briefe. Festschrift zu seinem 100. Geburtstag* (Leipzig: Hirzel 2005).

FIGURE 7.3 *The Dutch Nobel Prize winner Peter Debye poses for the sculptor Tjipke Visser, who is making a bust of the scientist in bronze by order of the Dutch Ministry of Education, Amsterdam 19 March 1937. Debye was professor of experimental physics in Leipzig between 1927 and 1934.*
PICTURE COURTESY OF THE NATIONAL ARCHIVES OF THE NETHERLANDS, THE HAGUE

opposed the new regime, but clearly held to their humanistic principles, and so courageously stood against it. The chemist Karl Friedrich Bonhoeffer, brother of Klaus and Dietrich Bonhoeffer, was one of them.[9]

After defeat in the Second World War and the end of National Socialism, the University of Leipzig faced a difficult new beginning. Over 60 per cent of the university buildings had been destroyed, many lecturers had been killed in the war or were still imprisoned in war camps, whilst those staff members who were still active at the university but who had worked closely with the National Socialist regime had to suspend their teaching and research activities.

The take-over of Leipzig by Soviet troops in June 1945 enabled the Soviet occupation and supported the political ambitions of the German communists. Immediately, the communists also tried to increase their influence on the university. This was achieved by creating new subfaculties, such as the Workers and Farmers faculty (*Arbeiter und Bauern Fakultät*) within the social

9 See pp. 210–215.

science faculty. The social science faculty itself was of great importance for the ideological saturation of the University of Leipzig. Although it formed an integral part of the university, it was governed by the central educational administration (*Deutschen Zentralverwaltung für Volksbildung*) in Berlin. Only politically reliable professors were allowed to work in this faculty. However, as part of the so-called second university reform in 1951, the social science faculty was closed and all its professors were amalgamated into the other faculties, particularly in the philosophical faculty. Through this reform as well as through the third university reform in 1968, the communist party of the German Democratic Republic, the Socialist Unity Party of Germany, succeeded in conclusively implementing its political and structural leadership at the University of Leipzig.[10] At the same time, there were plans to rebuild the university. The look of it had to be changed, both externally and internally. In 1953, it was renamed Karl Marx University and in 1968 the plans for the building works were finalised. As a consequence, a barbaric act was committed: the university church St. Pauli, which had survived the war undamaged, and the still functional old main building, the Augusteum, were completely destroyed.[11]

The political and ideological values of the communist party had such a strong influence at the university that up to the 1980s there was very little opposition. Only a few professors, such as the church historian Kurt Nowak, took part in the political and social opposition movement in 1989. The administration of the University of Leipzig on the other hand supported everyone who was prepared to use military means to crush the political opposition.[12]

The upheaval in 1989 enabled another reformation of the university. The resigning university council elected a new academic council in 1991 and decided to discard the name Karl Marx University and to revert to the old name University of Leipzig. In line with this change, the ethos, staff and

10 With regard to research about the history of the University of Leipzig in the first half of the twentieth century, see Ulrich von Hehl, "Zum Stand der Leipziger Universitäts- und Wissenschaftsgeschichte zur ersten Hälfte des 20. Jahrhunderts", in: Idem (ed.), *Sachsens Landesuniversität in Monarchie, Republik und Diktatur. Beiträge zur Geschichte der Universität Leipzig vom Kaiserreich bis zur Auflösung des Landes Sachsen 1952* (Beiträge zur Leipziger Universitäts- und Wissenschaftsgeschichte A, 03) (Leipzig: Evangelische Verlagsanstalt 2005): 19–50.

11 Christian Winter, *Gewalt gegen Geschichte. Der Weg zur Sprengung der Universitätskirche Leipzig* (Leipzig: Evangelische Verlagsanstalt 1998).

12 Michael Lippold, "Die Sektion Theologie an der Karl-Marx-Universität Leipzig im Herbst 1989. Die 'Wende' aus der Sicht eines damaligen Theologiestudenten", in: Andreas Gößner

structural setup of the university had to be reformed as well. Damaging disciplines such as Marxist-Leninist philosophy, pedagogy, journalism, law, history of the Soviet Union and military medicine were dissolved. At the same time, new areas of science were introduced and other existing high schools (*Hochschulen*) in Leipzig were amalgamated into the university. Many scientists from outside, especially from West Germany, were attracted by the prospect of putting the transition into practice. In result of these measures, the University of Leipzig has developed a new, strong performance profile since the mid 1990s – to become a modern, cosmopolitan and future oriented university.[13]

The Commission for the Study of the History of the University of Leipzig

At the end of the 1990s, a plan was conceived to write a new history of the University of Leipzig on the occasion of the upcoming 600th anniversary in 2009. Immediately the idea received the support from the former vice-chancellor, Cornelius Weiss, and former chancellor, Peter Gutjahr-Löser. The succeeding vice-chancellor Volker Bigl and the former pro-chancellor Günther Wartenberg put forward more concrete proposals for its conception. An interdisciplinary working group was established in order to promote the form and content of the university history. The group consisted of theologians, medical scientists, jurists, physicists, historians, art historians, educational scientists and archivists. The church historian Wartenberg took the post of chairperson. The working group quickly made progress and in 2002 it received the status of senate commission. Since then, it has been known as *Senatskommission zur Erforschung der Leipziger Universitäts- und Wissenschaftsgeschichte*.

Originally, the commission was financially supported by the Saxon Department for Science and Art (*Sächsischen Staatsministeriums für Wissenschaft und Kunst*), thus making it possible for the creation of the post of coordinator

(ed.), *Die Theologische Fakultät der Universität Leipzig. Personen, Profile und Perspektiven aus sechs Jahrhunderten Fakultätsgeschichte* (Beiträge zur Leipziger Universitäts- und Wissenschaftsgeschichte A, 02) (Leipzig: Evangelische Verlagsanstalt 2005): 461–468.

13 Peer Pasternack, *'Demokratische Erneuerung'. Eine universitätsgeschichtliche Untersuchung des ostdeutschen Hochschulumbaus 1989–1995. Mit zwei Fallstudien: Universität Leipzig und Humboldt-Universität zu Berlin* (Weinheim: Deutscher Studien Verlag 1999).

in 2002. He became responsible for overseeing the work undertaken in the different faculties and departments, the exchange of information between the faculty delegates and the distribution of the work among the authors. In 2007, a new staff member was employed to take over the overall compilation of the five-volume university history. For this purpose, Uwe John, a historian with a large amount of experience in producing multi-volume works, was appointed. At the same time, because the coordinator was extremely occupied with all kinds of tasks, another additional staff member was given the specific task of compiling a bibliography of books and articles with regard to the history of the university. This extensive bibliography developed into a basic tool for the authors in the faculties and departments. It was continually updated until the end of 2009.

In 2007, Wartenberg, the chairperson of the commission, suddenly died, triggering a crisis in the work of the commission. After all, he was an authority within the University of Leipzig as a whole and with regard to its history in particular and he would have been the author of a substantial number of chapters of the university's history. Moreover, Wartenberg, being a former pro-chancellor of the university, had excellent contacts in all circles of the university. It was only through his contacts that almost one hundred authors could have been recruited in order to write "The History of the University of Leipzig 1409–2009".

The members of the commission agreed to maintain continuity and to press ahead with the work that had already been started. The vice-chancellor Franz Häuser called forward the vice-chairperson of the commission, the historian Manfred Rudersdorf, to succeed Wartenberg as chairperson. The second vice-chairperson, the historian Ulrich von Hehl, continued with his duties. Probably even more important however, was to solve the question of who should take over the writing of the chapters that were designated to Wartenberg. The solution came quicker than expected. The church historian Klaus Fitschen from the faculty of theology quickly agreed to take over Wartenberg's task. With his engagement, he clearly helped the project out of an emergency situation.[14]

---

14 To the work of the commission, compares: Manfred Rudersdorf, "Universitätsgeschichte im Jubiläumsjahr. 600 Jahre Alma Mater Lipsiensis 1409–2009", in: *Leipziger Universitätsreden. Neue Folge Heft 108. Reden zur feierlichen Übergabe der Leipziger Rektoratsreden 1871–1933* (Leizig: Universitätsverlag 2009): 27–40; Ulrich von Hehl and Manfred Rudersdorf, "Vorwort", in: Bünz, Rudersdorf and Döring, *Geschichte der Universität Leipzig 1409–2009, Bd. 1* (2009): 11–20.

The History of the University of Leipzig in Five Volumes

The members of the working group agreed upon a concept for a multi-volume university history after long, sometimes controversial debates. Three aspects were to be especially considered. Firstly, the general development on the faculty and departmental level had to be described. Secondly, in addition, the scientific and structural development of the faculties and departments had to be arranged in their historical context. And thirdly, the development of humanities and sciences in Leipzig had to be put in a national and international perspective. The members of the working group were aware of the methodological problems of writing a university history. It was clear that in the twenty-first century, the writing of a university history was not to be seen as an antiquated tradition, being part of the jubilee ceremony. Instead it had to be a critical observation of the past that could also contribute to a better understanding of the present and its associated problems. Current standards should not be projected onto historical events. The art of historiography in this context is to fundamentally portray the differences and mutability of the institution in the past. In a time span of 600 years, very different organizational characteristics and institutional types hide under the name 'university'.

Regardless of the significance of the twentieth century for the development of Central European universities, university historiography must always hold global history in view. Along with the Christian churches, universities belong to the oldest still existing institutions of European history. In this respect, the university historian Anton Schindling from Tübingen concluded that universities have a significant dimension of continuity spanning several centuries. On the one hand, they share a number of common historical traditions and characteristics that are still sustained, related to the process of their foundation, traditional university names, old buildings and regalia. On the other hand, all individual universities indeed gradually acquired a specific character, but this was created through different answers on similar long-term and still relevant issues, such as the departmental system, the subject and faculty structure, examination and qualification criteria and teaching traditions whether or not in combination with science. On top of that, universities are generally considered to be part of the cultural heritage and therefore preserved, if only because of the large universal book collections in old university libraries.[15]

15 Anton Schindling, "Deutsche Universitäten in der Neuzeit: Eine Einführung in ihre Erforschung mit Würdigung der Arbeiten von Peter Baumgart", in: Peter Herde and Anton Schindling (eds.), *Universität Würzburg und Wissenschaft in der Neuzeit. Beiträge zur Bildungsgeschichte, gewidmet Peter Baumgart anlässlich seines 65. Geburtstags* (Würzburg:

However a university history also has to be critical towards all assumed continuity. It must be possible to avoid projecting ideas from the nineteenth century on to the early modern period. The historical development of universities is of importance for the current discussion of reforms in higher education, for example those related to the Bologna-process, but at the same time each period in history also has its own right without having to be relevant for later developments. Characteristic of the last 150 years is, for instance, to what extent the University of Leipzig has been influenced by extreme fractions. In that way the history of the university becomes a story of discontinuity and continuity, of impressive successes and achievement in its academic performance record in research and teaching, alongside dramatic decline.

All the publications with regard to the history of the university were composed according to these scientific and methodological guidelines. The first three volumes of "the history of the University of Leipzig" series offer a chronological overview of the structural and historical development.[16] The first volume is dedicated to the period 1409 to 1830/1831 and portrays the university in the context of the medieval and early modern history of Europe. The contributions were compiled by the historians Enno Bünz, Rudersdorf and Detlef Döring.[17] In the second volume, the classical university of the nineteenth century is presented. In this case, the authors are the historian Hartmut Zwahr and the archivists Gerald Wiemers and Jens Blecher.[18] The third volume deals with the period from 1909 until the present day. The historians Von Hehl and Günther Heydemann and the theologian Fitschen present the developments in the twentieth century, which were shaped by two dictators. The head of the department of human resources, Fritz König, gave an account of the structural and personnel changes at the university after 1990.[19]

Schöningh 1998): 15–35. Also compare in addition: Anton Schindling, *Bildung und Wissenschaft in der Frühen Neuzeit 1650–1800* (Enzyklopädie deutscher Geschichte 30) (München: Oldenbourg Verlag 1999); Notker Hammerstein, *Bildung und Wissenschaft vom 15. bis zum 17. Jahrhundert* (Enzyklopädie deutscher Geschichte 64) (München: Oldenbourg Verlag 2003).

16 The organization of the history of the University of Leipzig follows the classification pattern of Peter Moraw. Peter Moraw, "Aspekte und Dimensionen älterer deutscher Universitätsgeschichte", in: Peter Moraw and Volker Press (eds.), *Academia Gissensis. Beiträge zur älteren Gießener Universitätsgeschichte* (Marburg: Elwert 1982): 1–43.

17 Bünz, Rudersdorf and Döring, *Geschichte der Universität Leipzig 1409–2009, Bd. 1* (2009).

18 Hartmut Zwahr, Gerald Wiemers and Jens Blecher, *Geschichte der Universität Leipzig 1409–2009, Bd. 2: Das neunzehnte Jahrhundert 1830/31–1909* (Leipzig: Universitätsverlag 2011).

19 Ulrich von Hehl, Günther Heydemann, Klaus Fitschen and Fritz König, *Geschichte der Universität Leipzig 1409–2009, Bd. 3: Das zwanzigste Jahrhundert 1909–2009* (Leipzig: Universitätsverlag 2010).

The fourth volume deals with the history of individual faculties, disciplines and the traditional central facilities such as the university library, the art collection, the university archives and, somewhat exceptionally, also the department of German literature as being of importance for the university as a whole.[20] The structure of the volume follows the current faculty structure at the university. The authors at the different faculties were asked to write the history and development of their subject with the following central questions in mind:

- Firstly, the organizational-structural position of the respective scientific discipline was to be illustrated by taking into account the administrative as well as the philosophical and political conditions.
- Secondly, the specific paradigm shifts as well as the particular profile of the discipline at Leipzig University should be elucidated within the framework of the general scientific history of the subject.
- Thirdly, important people were to be presented, with regard to research as well as to teaching.

The contributions were organised in three different ways. In some cases, the faculty administration commissioned a few staff members to compile and write the history of the faculty. This happened in the faculties of theology, law, medicine, social science and educational sciences. Another option was that a team of authors in mutual agreement wrote the history of the faculty. This was the case for the faculties of chemistry and mineralogy. Thirdly, staff members of the individual departments wrote contributions with regard to their own scientific subject, which were then collated to compose a full history of the faculty. This method was considered necessary for faculties that were made up of scientific disciplines with different historical traditions, for example the faculty of biosciences, pharmacy and psychology and the faculty of history, art and oriental sciences. In result altogether, more than ninety authors from all faculties and departments contributed in one way or another to the fourth volume of the university's history. The presentation of these specific scientific developments in Leipzig is unique in Central European university historiography.

In the fifth volume, the history of the university in the urban context of the city of Leipzig from the Middle Ages to the current day is presented in an

20 Ulrich von Hehl, Uwe John and Manfred Rudersdorf (eds.), *Geschichte der Universität Leipzig 1409–2009, Bd. 4 (two half volumes): Fakultäten, Institute, Zentrale Einrichtungen* (Leipzig: Universitätsverlag 2009).

art-historical perspective. The volume is written by several art historians from the University of Leipzig under the leadership of Michaela Marek and Thomas Topfstedt. The volume is composed of two parts: a chronological and a systematical one. In the chronological part, the building and architectural history from the Middle Ages to the present construction schemes is presented. Additionally, the volume contains a catalogue of all the buildings that were used by and/or built for the university during its 600-year-old history (see figure 7.4). Also this work is unique within Central European university historiography.[21]

The project as a whole was commissioned by the vice-chancellor of the university and produced under the authority of the commission. The exception in this regard is volume four, the history of the faculties and departments, for which the different authors together with the deans and the school directorates are responsible themselves. All the authors of the five-volume university

FIGURE 7.4 *University of Leipzig and Mendebrunnen around 1900.*
PICTURE COURTESY OF LIBRARY OF CONGRESS, SEE WWW.LOC.GOV/PICTURES/ITEM/2002720592/(DATE ACCESSED 19/06/2014)

21 Michaela Marek and Thomas Topfstedt (eds.), *Geschichte der Universität Leipzig 1409–2009, Bd. 5: Geschichte der Leipziger Universitätsbauten im urbanen Kontext* (Leipzig: Universitätsverlag 2009).

history agreed to write scientific but still easily readable texts. The scientific traditions of the different subjects are reflected in the different styles of writing to be found in the history of the University of Leipzig. This shows how diverse the academic culture at the university is today. Besides, a uniform approach and style throughout the five-volume university history was admittedly not realistic.

In order to compile the history of the University of Leipzig, the collaboration of numerous institutions was necessary. Of particular importance was the collaboration with the university archives, the art collection and the library. In addition, the commission and individual authors cooperated with the city archives of Leipzig, the *Stadtgeschichtliches Museum Leipzig*, the city library, the *Sächsisches Staatsarchiv* in Leipzig, the main *Sächsisches Staatsarchiv* in Dresden and the German national library in Leipzig. Often illustrations were also acquired from further afield, e.g. from numerous national and international museums, libraries and archives.

Of course, it was not possible to tackle the extensive scientific and editorial work with the few members of the commission. Therefore, since the end of the 1990s, the historical institute of the University of Leipzig proposed scores of master and doctoral dissertations with regard to the theme of university history. To enhance the visibility of the results of that kind of research, the commission published many of these dissertations in a special series. Until now, twenty-four volumes of the series, titled *Beiträge zur Leipziger Universitäts- und Wissenschaftsgeschichte*, have been published. Some of them particularly proved their usefulness as preparatory studies for the five-volume university history, for instance *Sachsens Landesuniversität in Monarchie, Republik und Diktatur* edited by Von Hehl, and *Universitätsgeschichte als Landesgeschichte* edited by Döring.

The University of Leipzig provided funds for student and scientific assistants in order to support the personnel. They helped most of all with the work of the authors of volumes one to three. In 2007, the commission additionally sought the cooperation of the Leipzig job centre (*Leipziger Agentur für Arbeit*). Through one of their special support programmes, seven scientific projects involving fourteen staff members could be financed. After the actual scientific research was completed, the support staff also assisted with the editing work of the five-volume university history.

In addition to the activities of the commission, much other research with regard to the history of the University of Leipzig was initiated at the same time. In 2006, the head of the university library, Ulrich Johannes Schneider, started a project funded by the *Deutschen Forschungsgemeinschaft*. By publishing lecture catalogues of the University of Leipzig from 1815 to 1914 in a digital form, the aim was to create a tool for researching the history of individual disciplines and the specific profile of academic teaching in Leipzig.[22] Lecturers from the department of modern history set up a digital database, accessible on the internet, with biographical information of all the professors who have been teaching at the university, indeed at the moment still with a focus on the nineteenth century.[23] A similar project was undertaken in the department of church history. A lexicon of all professors and lecturers of the faculty of theology from 1409 to 2009 was created and published as volume A 8 of the series *Beiträge zur Leipziger Universitäts- und Wissenschaftsgeschichte*.[24] At the faculty of chemistry and mineralogy, a team of authors published an extensive history of the faculty in 400 pages next to their (shorter) contribution to volume four of the university history.[25]

In spite of all the endeavours of the university administration, the commission and all the authors, they unfortunately did not succeed in completing the five-volume university history on time for the jubilee in December 2009. Due to economical, technical and personnel reasons, another year and a half was required to finish the commission's work. During the jubilee year volumes four and five were published. In March 2010, the first volume was presented at the Leipzig Book Fair, in the autumn of the same year the third volume came out and volume two was ready in time for the Leipzig Book Fair of 2011.

After having finished his work as chairperson of the commission, Rudersdorf submitted a report to the senate of the university in which he not only gave a summary of the scientific and methodological approach, but also of the practical work of the commission. However, his intention to pursue the work of the commission in order to facilitate continuous research on university history could not be realized. Although Rudersdorf and Blecher (head of the university archives) were charged to initiate scientific research on the history of the University of Leipzig further, based on Rudersdorfs own report the work of the commission was declared completed and therefore it could be dissolved, thus destroying all structures regarding both university history and university education, which had been built up at the University of Leipzig over a period

---

22 See histvv.uni-leipzig.de/ (date accessed 19/06/2014).

23 See www.uni-leipzig.de/unigeschichte /professorenkatalog/ (date accessed 19/06/2014). See p. 9.

24 Markus Hein and Helmar Junghans (eds.), *Die Professoren und Dozenten der Theologischen Fakultät der Universität Leipzig von 1409 bis 2009* (Beiträge zur Leipziger Universitäts- und Wissenschaftsgeschichte A 8) (Leipzig: Evangelische Verlagsanstalt 2009).

25 Lothar Beyer, Joachim Reinhold and Horst Wilde (eds.), *Chemie an der Universität Leipzig. Von den Anfängen bis zur Gegenwart* (Leipzig: Passage-Verlag 2009).

of more than ten years. Although this might correspond with experiences at other universities (e.g. at the University of Oslo), it casts a damning light on those universities. Renouncing continuous research on university history will surely prove to be a striking mistake of the latest university jubilee.

In result, it was not the commission to make the final point, but the former vice-chancellor Häuser. He assembled a team around him, composed of the rector's consultant Michael Handschuh, the head of the university archives and the coordinators of the commission Jonas Flöter and Sebastian Kusche. Together they published a collection of documents with regard to the events taking place at the University of Leipzig in the jubilee year, dealing with conferences, symposia and publications. In the book the appreciation is expressed for the commitment of all commissions and task groups working on the university jubilee and, as an attachment, it contains numerous pictures commemorating the jubilee celebrations.[26] The book makes clear to what extent the University of Leipzig owns a solid basic concept of its self-assurance, also thanks to its five-volume history. With this collection Leipzig is one of the few German universities offering a recent, detailed overall view of its own history to a broad public.

26 Franz Häuser (ed.), *Das sechshundertjährige Jubiläum der Universität 2009. Eine Dokumentation* (Leipzig: Universitätsverlag 2011).

PART 3

University History Writing Beyond the Jubilee

CHAPTER 8

The Humboldtian Tradition

The German University Transformed, 1800–1945

Johan Östling

Wilhelm von Humboldt is often hailed as the father of the modern research university. The essentials of his university programme are familiar: a combination of research and teaching; Lehr- und Lernfreiheit; Bildung *rather than vocational training; a vision of the unity of science. Similarly famous is the significance of the University of Berlin (founded in 1810) as the consummate seat of learning in Germany and beyond.*

In recent decades, however, historians have set out to revise this picture. Firstly, the originality of Humboldt's ideas has been questioned, especially in the light of the influence exerted by other neo-humanist philosophers. Secondly, it has been noted that there was a long hiatus before the model had any impact outside Berlin. At the same time, a more thorough historicizing of the Humboldtian tradition has begun. In effect, the point made in the new research is that 'Humboldt' was not born in 1810, but in 1910 when the University of Berlin celebrated its centenary.

The rediscovery of Humboldt in the early twentieth century must be seen in the context of general developments in the German Empire in the decades running up to the First World War. It was a dynamic period, when science, education, and history became essential components in the cultural identity then taking shape in the young German nation-state. In the wider perspective, it is clear that the Humboldtian legacy has always been refashioned to reflect the societal interests, historical experiences, and national aspirations of the day.

This article brings together important insights drawn from the most recent research, and discusses the main features of the Humboldtian tradition's emergence and transformation. Moreover, it gives a general historical background to the development of an academic self-understanding, which not least has been manifested at various university jubilees. The emphasis is on the decades around 1800 and 1900 respectively, although developments in the interwar period are considered as well.

The Humboldtian Tradition

When the University of Berlin opened its doors in October 1810, a new era in the history of higher education dawned. The man behind this achievement

 | DOI 10.1163/9789004265073_009

was the Prussian government functionary, linguist, and educational reformer Wilhelm von Humboldt. In an epoch-making, programmatic essay, he had drawn up the guidelines for a new seat of learning where education and scholarship were to be more important than the rote learning of received truths; where there was to be a dynamic link between the production and communication of knowledge; where academic freedom was to have a wider, more certain meaning. In short, the modern university had seen the light of day.

This in a nutshell is the view of modern university history that dominated in the twentieth century, a picture of the past that has been the basis of academic self-understanding and has underpinned intellectual identity. Even today, perhaps especially today, Humboldt's name is on the lips of keynote speakers and invoked in debate whenever the future of the university is raised,[1] not least when it comes to his mother country, Germany. At the same time, the past two decades have seen a thorough historicizing of Wilhelm von Humboldt's work and the academic ideals associated with his name. Scholars have questioned his originality, and have pointed out that by no means all of his ideals were ultimately realized in the university that today bears his name, the *Humboldt-Universität zu Berlin*. They have shown that Humboldt was almost wholly absent from nineteenth-century discussions of universities, education, and knowledge. His works were unknown; his ideas barely considered. The universities of Halle and Göttingen, rather than their Prussian counterpart in Berlin, were the benchmarks of the day. However, at the turn of the twentieth century Humboldt was suddenly rediscovered. His university plan was finally published and became famous when the University of Berlin celebrated its centenary. At the same time, influential educational politicians and educationalists spread his ideas. The university ideologist Wilhelm von Humboldt was born – posthumously, not in 1810 but in 1910.

Despite the belated interest, it cannot be denied that the Humboldtian tradition has been an important academic ideal in many of the major confrontations over higher education. For a century it has been a constant source of inspiration for critics, reformers, and evokers of past glories. Thus, the new research does not require a denial of the Humboldtian tradition, even less a belittling of its significance; instead it approaches it with a deeper understanding of the role that it has played – and continues to play – in the conflict over university goals and meaning.

1 Thomas Karlsohn, Peter Josephson and Johan Östling (eds.), *The Humboldtian Tradition: Origins and Legacies* (Scientific and Learned Cultures and Their Institutions 12) (Leiden: Brill 2014).

In what follows, I amass significant insights from the recent decades of research and sketch the main features of the Humboldtian tradition's emergence and transformation. The purpose of this synthesis of the historiography is not only to discuss how the interpretations of a Prussian legacy have shifted in different contexts and epochs. In addition, the aim is to give a general, overarching historical background to the development of an academic self-understanding, which not least has been manifested at various university jubilees. For obvious reasons the main emphasis will be on the decades at the turn of the nineteenth and twentieth centuries, also because so much of the historiography has focussed on these two periods. I also outline the tendencies in the interwar period, but go no further than 1945. After the Second World War, the nature of the problems changed and research on the post-war period is much scarcer.[2]

Humboldt and His Age

In accounts of European university history, the eighteenth century is often characterized as a period of decline. Indeed, new seats of learning continued to be founded, but there were loud calls for the very structure of education to be reformed. There was a growing opinion that teaching should be reorganized and directed at the needs of professional life, rather than be allowed to continue to peddle old learning. However, reforming the universities was a drawn-out process, and during the eighteenth century a series of special schools were founded in subjects such as veterinary medicine, mining, and trade to meet the demands of the new age. It was at these colleges, as well as at learned societies, that the vast proportion of research would be conducted. Towards the end of the eighteenth century, the university as an institution was not held in particularly high esteem, be it by utilitarian natural scientists, rationalist Enlightenment philosophers, or the burgeoning middle classes. Together with the Church, the universities, with their medieval, confessional stamp, became the emblem of the Ancien Régime.[3]

2 In a current research project, pursued within the Pro Futura Scientia Programme and financed by *Riksbankens Jubileumsfond*, I trace the transformation of the Humboldtian tradition in post-war Germany. For a general overview of the developments since 1945, see Konrad H. Jarausch, "The Humboldt Syndrome. West German Universities, 1945–1989 – An Academic Sonderweg?", in: Mitchell G. Ash (ed.), *German Universities, Past and Future: Crisis or Renewal?* (Providence, RI: Berghahn 1997): 33–54.

3 Hilde de Ridder-Symoens (ed.), *A History of the University in Europe. Vol. 2 Universities in Early Modern Europe (1500–1800)* (Cambridge: University Press 1996): 52–80.

Recently, however, a number of historians have problematized this account of European universities. Several of them have suggested that the eighteenth-century enlightened monarchs were well aware of the stagnation and launched reforms to instil new life into higher education. This being the case, the developments of the nineteenth century become mostly the completion of a transformation that was already well underway.[4] Others have argued that the origins of the modern university must be understood as one strand in the development of the bureaucratic state. In an original and wide-ranging study, William Clark, for example, has argued that the growing state administration sought to limit the old, academic liberties and increase political control. At the same time, he maintains that the rise of the modern university must be seen in relation to the emerging market for books. When books became more widely available, university professors were exposed to competition as learned authorities. This had an impact on their role, and in the long run put a premium on groundbreaking research. And according to Clark, all these fundamental changes took place in the eighteenth century.[5]

Despite these new interpretations, it is hard to deny that the Enlightenment, the French Revolution, and the ensuing tumult contributed to the transformation of the academic system in Europe. In the wake of the Napoleonic Wars, two new principal models of higher education crystallized: the French and the Prussian. In many parts of Europe, much would remain unchanged for a long time yet, but in two large areas of the Continent events took a very different turn.[6] In France, the universities' autonomy came to an end, and instead they were subordinated to the power of the political regime. *Collèges* and traditional faculties were replaced by a series of vocational and special schools, although a number of older institutions, such as the *Collège de France*, did survive both the Revolution and Napoleon, and it was particularly at these institutions that a substantial proportion of French research was conducted. All in all,

---

4 Robert D. Anderson, *European Universities from the Enlightenment to 1914* (Oxford: University Press 2004): 4, 20–38 and 51–52.

5 William Clark, *Academic Charisma and the Origins of the Research University* (Chicago: University Press 2006).

6 Walter Rüegg, "Themes", in: Walter Rüegg (ed.), *A History of the University in Europe. Vol. 3 Universities in the Nineteenth and Early Twentieth Centuries (1800–1945)* (Cambridge: University Press 2004): 3–13; Christophe Charle, "Patterns", in: Rüegg (ed.), *A History of the University in Europe* (2004): 33–40; Anderson, *European Universities from the Enlightenment to 1914* (2004): 3–5 and 63–64; Gert Schubring, "Spezialschulmodell versus Universitätsmodell: Die Institutionalisierung von Forschung", in: Idem (ed.), *"Einsamkeit und Freiheit" neu besichtigt. Universitätsreformen und Disziplinenbildung in Preussen als Modell für Wissenschaftspolitik im Europa des 19. Jahrhunderts* (Stuttgart: Steiner Verlag 1991): 288–296.

as a result of the upheavals, the academic reality of nineteenth-century France was marked by specialization and fragmentization.[7]

In Prussia, however, the university both as an idea and an institution saw something of a renaissance. Already in the eighteenth century new features had been incorporated into academic life, in particular at those epitomes of the Enlightenment, Göttingen and Halle. One of these new characteristics, which would prove crucial, was the requirement that professors should not only teach but also conduct research. In addition, lectures had to be complemented with a seminar, being a forum for scholarly discussion between students and researchers–lecturers.[8] The old, but enduring, order of the faculties began to be questioned more and more at the end of the eighteenth century. At the medieval universities, the philosophical faculty was the lowest, mainly devoted to preparatory studies. However, Immanuel Kant, Johann Gottlieb Fichte and Friedrich Schleiermacher argued that the philosophical faculty ought to be the equal, or even the superior, of the other three faculties. In *Der Streit der Fakultäten* (1798), Kant wrote about the conflict between the philosophical faculty on the one hand and the theological, legal, and medical faculties on the other. Despite the accepted view, Kant contended that the philosophical faculty was the most advanced because it was free from expectations of utility and government meddling.[9]

Thus the promise of much that would blossom in the nineteenth century and become known as the distinguishing marks of the German university was already in place. A good many small reforms had already been pushed through, and the debate on academic ideals was already well underway in the latter eighteenth century. Meanwhile, the growth and establishment of a distinctively Prussian university model must be seen against the momentous events of the years around 1800.

The Revolutionary and Napoleonic Wars marked entire generations. Later research may have nuanced the picture somewhat, by emphasizing that not all regions were as severely hit by the French campaigns and occupation, but as far as Prussia was concerned the fact remains that the humiliating

7 Charle, "Patterns" (2004): 33–47; Schubring, "Spezialschulmodell versus Universitätsmodell" (1991).

8 Charle, "Patterns" (2004): 47–48; Roy Steven Turner, "University Reforms and Professorial Scholarship in Germany 1760–1806", in: Lawrence Stone (ed.), *The University in Society: Europe, Scotland, and the United States from the 16th to the 20th Century* (Princeton: University Press 1974), vol. 2: 495–532.

9 See also Riccardo Pozzo, "Kant's *Streit der Fakultäten* and Conditions at Königsberg", *History of Universities* 16 (2000), no. 2: 96–128.

set-backs – the capture of Berlin, the twin defeats of Jena and Auerstedt – prompted a strong reaction. Out of resistance against French ascendancy grew an aversion to the very cosmopolitanism of the Enlightenment. Bitter experience fuelled a patriotic awakening – a nascent German nationalism with Prussian overtones. At the same time, the defeats prompted a soul-searching that paved the way for some of many awaited reforms of important social institutions. Unlike in revolutionary France, the changes in Prussia were introduced gradually in the wake of the Napoleonic Wars, among them the abolition of serfdom, the emancipation of the Jews, free trade, and compulsory military service; all crucial in transforming Prussia from a feudal realm to a modern, industrial state.[10]

The foundation of the University of Berlin in 1810 cannot be explained without taking into account this political and social context. In 1789, thirty-five universities existed in the German area and almost half the students had matriculated at one of the four largest among them (Halle, Göttingen, Jena, or Leipzig). A quarter of a century later, only sixteen universities remained; the others had been abolished or forced to close their doors in the aftermath of war and conquest. Moreover, Prussia had lost its former academic flagship, when in 1807 the University of Halle became part of Napoleon's new creation, the kingdom of Westphalia. The truth of Frederick William III's words – "der Staat muss durch geistige Kräfte ersetzen, was er an physischen verloren hat [the state must replace with spiritual strength that which it has physically lost]" – has been questioned by historians,[11] yet even so, one cannot ignore the fact that the refashion of the Prussian educational system was not only an important element in a series of general reforms, but was also an attempt to meet the challenge of the Napoleonic special schools. The fact that two other universities, both duly named *Friedrich-Wilhelms-Universität*, were founded in Prussia at this point – in Breslau in 1811 and Bonn in 1818 – only enforces this impression.[12]

10 See, for example, Hans-Ulrich Wehler, *Deutsche Gesellschaftsgeschichte: Vom Feudalismus des Alten Reiches bis zur Defensiven Modernisierung der Reformära 1700–1815* (München: Beck 1987); David Blackbourn, *History of Germany, 1780–1918: The Long Nineteenth Century* (Malden: Blackwell 2003).

11 Rüdiger vom Bruch, "Die Gründung der Berliner Universität", in: Rainer Christoph Schwinges (ed.), *Humboldt international: Der Export des deutschen Universitätsmodells im 19. und 20. Jahrhundert* (Veröffentlichungen der Gesellschaft für Universitäts- und Wissenschaftsgeschichte 3) (Basel: Schwabe Verlag 2001): 58–60; Charle, "Patterns" (2004): 33–34.

12 Charles E. McClelland, *State, Society, and University in Germany, 1700–1914* (Cambridge: University Press 1980): 102–108; Schubring "Spezialschulmodell versus Universitätsmodell" (1991): 288–296.

Obviously, the University of Berlin did not spring from an intellectual vacuum. Since the latter eighteenth century, the idea of the university had been dealt with in numerous treatises and debates: Kant's, Fichte's and Schleiermacher's observations on the position of the philosophical faculty came into existence as part of this discussion, but at the same time had far greater implications. A vital centre for the exchange of ideas was Jena in Thuringia. The city's university had been founded back in the mid-sixteenth century, and by the end of the eighteenth century it had developed into a rare academic free zone. In the period around the 1790s, the city assembled all the leading thinkers of the day as professors at the university, including Friedrich Schiller, Fichte, Schelling and August Wilhelm Schlegel. Even more important for the creative atmosphere was the stream of authors, artists, and philosophers who came to Jena for varying periods: Goethe, Hölderlin, Novalis, Schleiermacher, Caroline Schlegel and Dorothea Veit. This was the milieu that inspired hopes for a new kind of educational institution, "the Romantic university", with education, academic freedom, and collective research as its cornerstones.[13]

Some have described the University of Berlin as being the "institutionalization of the Jena ideal".[14] To a certain extent this makes sense, but on the other hand the new university that gradually took shape also had its own specific antecedents. After all, already from 1784, it had been proposed to establish a new university in the city of Berlin. During the first decade of the nineteenth century, many of the ideas that had been in circulation in Jena were developed further, and thinkers such as Schelling, Fichte and Steffens all penned the outline of a new university, with education and pure scholarship as its guiding principles. In the majority of these draft reforms of higher education, the

---

13 Gerhard Müller, Klaus Ries and Paul Ziche (eds.), *Die Universität Jena: Tradition und Innovation um 1800* (Stuttgart: Steiner Verlag 2001); Thomas Karlsohn, "Det romantiska universitetet", *Psykoanalytisk Tid/Skrift* 26–27 (2009): 97–123.

14 Theodore Ziolkowski, *German Romanticism and Its Institutions* (Princeton: University Press 1990): 286. As Thorsten Nybom sees it, the other great university ideologist in the nineteenth century, Cardinal Newman, took an existing institution – Oxford – as his model when he wrote *The Idea of a University*, whereas Humboldt drew on a number of existing ideas and incorporated them in one institution, the University of Berlin. See Thorsten Nybom, "A Rule-Governed Community of Scholars: The Humboldt Vision in the History of the European University", in: Peter A.M. Maassen and Johan P. Olsen (eds.), *University Dynamics and European Integration* (Dordrecht: Springer 2007): 79. For a more general view on Newman as a university reformer, see Sheldon Rothblatt, *The Modern University and Its Discontents: The Fate of Newman's Legacies in Britain and America* (Cambridge: University Press 1997).

charged term "university" was shunned in favour of terms such as "institute of higher education" or "educational establishment". One noted exception was Schleiermacher's *Gelegentliche Gedanken über Universitäten in deutschem Sinn* (1808), the most significant and widely disseminated of the treatises to be published. Inspired by Romantic circles in Jena, he drew up the guidelines for a new type of university.[15]

In his treatise, Schleiermacher also was at pains to address another, more specific question. Why this new type of university had to be established in Berlin? In his opinion, other Prussian cities could indeed more readily attract students and lecturers than the expensive and relatively peripheral capital city, but Berlin had some obvious advantages, which these other cities were missing. It provided all kinds of institutions the new university could benefit from: a world-class library, an observatory, a menagerie and an anatomical theatre. The same applied to a number of special schools that had been set up along the river Spree.[16]

In July of the following year, 1809, Frederick William III of Prussia received a document that presented a similar argument. The author, who had clearly been influenced by Schleiermacher, recommended that a public seat of learning had to be founded. A telling argument in favour of locating it in Berlin was the existence of institutions, collections, and academies in the city, inextricably bound up with the fact that they in turn could not do themselves full justice if they were not linked to the new institution's research-based teaching. The idea was that all institutions would retain their independence, but that at the same time they would be heavily dependent on one another. The memorandum was signed by Wilhelm von Humboldt.[17]

Humboldt was then the government minister responsible for Prussia's cultural and educational affairs, but his career until that point had been a rather varied one. He was born in Potsdam on 22 June 1767 as the eldest son of Alexander Georg von Humboldt, a royal chamberlain, and his wife Marie-Elisabeth von Colomb. Together with his younger brother Alexander, later a world-famous naturalist, he received a thorough and, for the nobility,

15 Vom Bruch, "Die Gründung der Berliner Universität" (2001): 57–62. Many of the tracts on the university or higher education from the early nineteenth century have been published in Ernst Anrich (ed.), *Die Idee der deutschen Universität: Die fünf Grundschriften aus der Zeit ihrer Neubegründung durch klassischen Idealismus und romantischen Realismus* (Darmstadt: Wissenschaftliche Buchgesellschaft 1956) and Ernst Müller (ed.), *Gelegentliche Gedanken über Universitäten* (Leipzig: Reclam 1990).

16 Vom Bruch, "Die Gründung der Berliner Universität" (2001): 59.

17 Vom Bruch, "Die Gründung der Berliner Universität" (2001): 60.

typical education provided by leading private tutors. In 1787, he matriculated at the Alma Mater Viadrina in Frankfurt (Oder), where he stayed only for a year, eager to move to the livelier city and university of Göttingen, where the combination of learning and Enlightenment ideas made a lasting impression on him. After four terms of study he had toured Western Europe, experienced the revolutionary fervour of Paris, become engaged to Karoline von Dacheröden, and struck up a correspondence with Schiller, Goethe and Herder, before joining Prussian government service in 1790. Already after approximately a year, he resigned from his post, though fully determined to dedicate himself to his studies. Between 1793 and 1797, he lived in Jena, where he enjoyed a few immensely productive years in the company of Schiller, Fichte and the Schlegel brothers, cultivating his philosophical and philological interests. The Jena-period was followed by a longer sojourn in Romance-speaking Europe, primarily based in Paris, but with several longer journeys to Spain and other countries. After a short interlude in Berlin, Humboldt spent the years 1802–1808 as a Prussian diplomat to the Holy See, which was in many ways the richest period of his life, when he and his wife could daily enjoy Rome's antiquities and mix with the artistic set. Thereafter Humboldt was recalled to the Prussian capital and appointed minister in vom Stein's reform cabinet. As a result of his sixteen prolific months' service, from February 1809 to June 1810, Humboldt was to leave an abiding mark on the educational system.[18]

His efforts as a government minister were initially directed at shaking up the Prussian school system. His educational vision took in all phases of education – from primary school, through the gymnasium, right up until the university. In the summer of 1809, he therefore presented his paper on the founding of a new university in Berlin to Frederick William III, and in August of that year the King formally approved the proposal. It took a bit

18 The literature on Wilhelm von Humboldt has been growing for more than half a century. The first full treatment was Rudolf Haym, *Wilhelm von Humboldt: Lebensbild und Charakteristik* (Berlin: Rudolf Gaertner 1856). Among the more recent, Paul R. Sweet, *Wilhelm von Humboldt: A Biography*, 2 vols. (Columbus: Ohio State University Press 1978–1980) is for its range and thoroughness in a class of its own. Likewise, Herbert Scurla, *Wilhelm von Humboldt: Werden und Wirken* (Düsseldorf: Claassen 1976) is a solid account, only somewhat marred by its Eastern European Marxist jargon. Several older, still much-cited works offer their own contributions without being biographies as such: Eduard Spranger, *Wilhelm von Humboldt und die Humanitätsidee* (Berlin: Reuther & Reichard 1909); Siegfried A. Kaehler, *Wilhelm von Humboldt und der Staat* (München: Oldenbourg Verlag 1927); Menze Clemens, *Die Bildungsreform Wilhelm von Humboldts* (Hanover: Schroedel 1975).

more than a year, until October 1810, before teaching and research could start at the Alma Mater Berolinensis, and even then it began on a small scale. Only after six years, for instance, real statutes were put in place. When the *Georg-August-Universität* had opened its doors in Göttingen in 1737, it had given occasion to an extravagant inaugural ceremony. In Berlin however, which was scarred by defeat and years of famine, the new university was welcomed without pomp.[19] Nevertheless, it soon established its academic reputation, largely thanks to Humboldt's success in persuading so many exceptionally gifted scholars to accept a chair. Fichte was to be the first holder of the ever-important chair of philosophy (including a spell as rector of the university) and was succeeded in 1818 by Hegel; Schleiermacher was the first to be appointed to the chair of theology; Carl von Savigny in law; Carl Ritter in geography; August Böckh in classical philology; and Christoph Wilhelm Hufeland in medicine. Gradually, a series of leading academic circles were established, Leopold von Ranke's history seminar and Johannes Müller's physiological research group among them. But by this point, Wilhelm von Humboldt had long since ceased to be in charge of the Prussian universities.[20]

To conclude, the new University of Berlin was thus not the work of a moment. On the contrary, it was preceded by the long reform processes of the eighteenth century, the intense intellectual debate in the years around 1800, and the political reaction to Prussia's defeat by Napoleonic France. From out of all this, a new university system was born, although for many years it was more an idea than a reality. Yet it is precisely the idea that is the central interest of our analysis.

---

19 Vom Bruch, "Die Gründung der Berliner Universität" (2001): 53–54.

20 A new, scholarly history of the University of Berlin was planned on the occasion of the bicentenary in 2010, *Geschichte der Universität Unter den Linden 1810–2010* (Berlin: Akademie Verlag), and the six volumes, some of them somewhat belated, have been published in recent years: Heinz-Elmar Tenorth and Charles McClelland (eds.), *Vol. 1: Gründung und Blütezeit der Universität zu Berlin 1810–1918* (2013); Heinz-Elmar Tenorth and Michael Grüttner (eds.), *Vol. 2: Die Berliner Universität zwischen den Weltkriegen 1918–1945* (2012); Konrad H. Jarausch, Matthias Middell and Annette Vogt (eds.), *Vol. 3: Sozialistisches Experiment und Erneuerung in der Demokratie – die Humboldt-Universität zu Berlin 1945–2010* (2012); Heinz-Elmar Tenorth (ed.), *Vol. 4: Genese der Disziplinen: Die Konstitution der Universität* (2010); Heinz-Elmar Tenorth (ed.), *Vol. 5: Transformation der Wissensordnung* (2010); Heinz-Elmar Tenorth (ed.), *Vol. 6: Selbstbehauptung einer Vision* (2010).

Humboldt's Ideas

As a university ideologist, Wilhelm von Humboldt did not swim against the currents of the day. He was a skilful synthesizer who achieved success by systematically mustering ideas that were already in circulation and by giving his arguments the most pregnant expression. His ministerial tenure, however brief, gave weight to his words. It could be claimed that Humboldt succeeded in turning reform directed against the university into reform of the university.

More than anything else, it was the term *Bildung* that came to be associated with Humboldt's name. And not without reason; the term was wholly central to his educational philosophy. The word itself is very old, related to *Bild*, *Abbildung*, and many other derivations, and throughout the centuries its meaning broadened successively, so that by the eighteenth century it had generally acquired the meaning of 'formation'. In the decades around 1800, the word was taken up in public debate and started to gain currency within the growing middle class (*Bildungsbürgertum*). Even if there were obvious shades of meaning, it came to be the common source of the specific union of neo-humanism, Enlightenment and idealism that was so characteristic of the intellectual climate of German-speaking Europe. It is telling that *Bildung* has no direct equivalent in the other major European languages. Translations such as *éducation*, 'formation', or 'self-cultivation' fail to capture the full range of meaning of the German word.[21]

As an educational concept, the German *Bildung* is close kin with other, far older terms. It can be traced back to the Greek *paideia*, the first programme for the all-round development of man's spiritual, aesthetic, and physical abilities with the intent of forming a whole, harmonious citizen. Likewise, the concept of education that emerged in the eighteenth century was inspired by a late medieval interpretation of the old Christian precept that man should strive to become the image of God, *Imago Dei*.[22] The vestiges of this idea were to be found in the

21 Rudolf Vierhaus, "Bildung", in: Otto Brunner, Werner Conze and Reinhart Koselleck (eds.), *Geschichtliche Grundbegriffe: Historisches Lexikon zur politisch-sozialen Sprache in Deutschland* (Stuttgart: Klett 1972): 508–551; Ernst Lichtenstein, "Bildung", in: Joachim Ritter (ed.), *Historisches Wörterbuch der Philosophie* (Basel: Schwabe Verlag 1971): 921–927; Reinhart Koselleck, "Einleitung – Zur anthropologischen und semantischen Struktur der Bildung", in: Idem (ed.), *Bildungsbürgertum im 19. Jahrhundert* (Stuttgart: Klett-Cotta 1990): 11–46.

22 Vierhaus, "Bildung" (1972); Liechtenstein, "Bildung" (1971). The classic work on education in classical antiquity is Werner Jaeger, *Paideia: Die Formung des griechischen Menschen*, 3 vols. (Berlin: W. de Gruyter 1934–1947).

work of influential educational theorists such as Johann Gottfried Herder, who was one of the first to formulate a more coherent educational vision with *Bildung* at its heart, the overall purpose of which was to develop the individual's abilities, eschewing the premium placed on the mechanical learning of an agreed curriculum. Many of the great names of the day – Johann Wolfgang von Goethe, Friedrich von Schiller, Johann Heinrich Pestalozzi, Immanuel Kant – referred to Herder and contributed to the dynamic, late eighteenth- and early nineteenth-century discussions of educational theory and practice.[23]

Even so, it was only through Wilhelm von Humboldt that the idea of *Bildung* was truly integrated into an educational programme and put on an institutional footing. In line with his contemporaries, for Humboldt *Bildung* was about the furthest and most harmonious development of man's inherent abilities. His theoretical arguments about education show evidence of a kind of duality in this respect. On the one hand he describes an educational process in which the unfettered refinement of character was central – his *Bildung* rested on the subjective assimilation of knowledge that both originated in and would transform the individual. On the other hand, the development of the individual always stood in relation to history and everything that was truly human. The realization of man's inner potential subsisted in a dialectical movement between the ego and the surrounding culture.[24]

Humboldt, however, was as practical as he was theoretical, and he made his educational views concrete in several proposals, memoranda, and drafts, which he wrote during his time as a government minister. An official document of 1809, for instance, outlines an educational system that would provide pupils of a *Menschenbildung*. Teaching was neither to be narrowly specific nor strictly vocational; instead pupils would concentrate on what was truly human, on "the main forces of the intellect". Humboldt consistently emphasized the importance of wide-ranging studies in languages, history, and mathematics, but he also had been strongly influenced by neo-humanism. Classical subjects, primarily the Greek curriculum, held a natural and unique position for

23 Koselleck, "Einleitung – Zur anthropologischen und semantischen Struktur der Bildung" (1990).

24 Wilhelm von Humboldt, "Theorie der Bildung des Menschen: Bruchstücke", in: Idem, *Werke in Fünf Bänden. Vol. 1 Schriften zur Anthropologie und Geschichte* (Darmstadt: Wissenschaftliche Buchgesellschaft 2010): 234–240. The literature on Wilhelm von Humboldt's theory of education is extensive. See Dietrich Benner, *Wilhelm von Humboldts Bildungstheorie: Eine problemgeschichtliche Studie zum Begründungszusammenhang neuzeitlicher Bildungsreform* (Weinheim: Juventa 2003).

him.[25] In another official document, also published in 1809, Humboldt set out the guidelines for the examination of Prussian government officials. It reveals that Humboldt's vision of education was not narrowly limited to schools. He argued that it was immaterial to future civil servants whether they could rehearse statistics or individual facts. Rather, it was the man's intellectual vitality and general ideas about humanity that would determine how qualified he was. Therefore, these qualities and character traits should be tested in order to judge a man's fitness to enter the service of the state.[26]

Thus, *Bildung* did not only hold a manifestly central place in Humboldt's educational philosophy, it was also fundamental to his idea of the university. His academic vision was articulated most clearly in *Über die innere und äussere Organisation der höheren wissenschaftlichen Anstalten in Berlin*. This short, unfinished paper, written at the end of 1809 or in the spring of 1810, shows many of the key ideas that woul later be synonymous with the Humboldtian tradition.[27]

Alongside *Bildung*, the notion of *Wissenschaft* (science or scholarship) was a cornerstone in Humboldt's concept of the university.[28] In the 1809/1810 paper a self-evident link was drawn between them. Humboldt stated that the university should be the place where scholarship in its deepest, widest, and purest sense should reign. "Da diese Anstalten ihren Zweck indess nur erreichen können, wenn jede, soviel als immer möglich, der reinen Idee der Wissenschaft gegenübersteht, so sind Einsamkeit und Freiheit die in ihrem Kreise vorwaltenden Principien," Humboldt asserted. Unlike primary and secondary schools, which existed to pass on fixed, fully-fledged knowledge to the children, scholarship should be seen as "ein noch nicht ganz aufgelöstes Problem". The university would stand or fall on the defence of the principle that *Wissenschaft* ought to be seen – as the key formulation put it – as "etwas noch nicht ganz Gefundenes und nie ganz Aufzufindendes". Humboldt stuck to his educational ideal when he asserted that it is only the scholarship,

25 Wilhelm von Humboldt, "Der Königsberger und der Litauische Schulplan", in: Idem, *Werke in Fünf Bänden. Vol. 4 Schriften zur Politik und zum Bildungswesen* (Darmstadt: Wissenschaftliche Buchgesellschaft 2010): 168–195.

26 Wilhelm von Humboldt, "Gutachten über die Organisation der Ober-Examinations-Kommission", in: Idem, *Werke in Fünf Bänden* (2010), vol. 4: 77–89.

27 Wilhelm von Humboldt, "Über die innere und äussere Organisation der höheren wissenschaftlichen Anstalten in Berlin", in: Idem, *Werke in Fünf Bänden* (2010), vol. 4: 255–266.

28 In German minds, *Wissenschaft* of course comprises both human and natural sciences. Fritz K. Ringer, *The Decline of the German Mandarins: The German Academic Community, 1890–1933* (Cambridge: Harvard University Press 1969): 102–104.

which is derived from the inner being, that could in turn shape the character, and that it must be both the state's and mankind's goal to generate character and action. To this end, it is essential that it all derives from one idea, and that all forms of narrow-mindedness are discouraged.[29]

Humboldt also advanced the theory of academic freedom. The state should not treat its university as a gymnasium or a special school, nor use it as a store for experts that might come in handy. On the contrary, the state could not expect anything of the academy that had immediate bearing on the state itself. The state's principal duty was to ensure that its secondary schools best served the higher educational institutions. If the latter were instituted and run in an ideal manner, the pupils would develop a desire to dedicate themselves to scholarship.[30]

Towards the end of his text, Humboldt argued against the notion that the university ought to engage itself only in teaching, and that research should be conducted at special academies. If scholarship is not seen to be changeable, then it is not worthy of being called scholarship, he concluded.[31]

Many current researchers point to this unfinished fragment from 1809/1810 as the key to understanding Humboldt's idea of the university. Already in 1993, Björn Wittrock called it "perhaps the most discussed document in the modern history of universities".[32] In this and a number of other treatises from the same period, one can discern the academic principles that would come to be conjured up when mentioning Humboldt's name: academic freedom, the coupling of teaching and research, the intellectual fellowship of lecturers and students, scholarship as education and the like. The Humboldtian tradition at the same time is so much richer and more nebulous. It cannot be narrowed down to a few points. Its transformation in the course of the two centuries that have elapsed since, reflect Germany's overwhelming history.

Humboldt in the Nineteenth Century

Wilhelm von Humboldt died on 8 April 1835. In the quarter of a century that had passed since he had handed over responsibility for the Prussian education

29 Humboldt, "Über die innere und äussere Organisation" (2010), vol. 4: 256–258.

30 Humboldt, "Über die innere und äussere Organisation" (2010), vol. 4: 260–261.

31 Humboldt, "Über die innere und äussere Organisation" (2010), vol. 4: 263–266.

32 Björn Wittrock, "The Modern University: The Three Transformations", in: Sheldon Rothblatt and Björn Wittrock (eds.), *The European and American University since 1800: Historical and Sociological Essays* (Cambridge: University Press 1993): 317.

system in 1810, he had dedicated himself to diplomacy and linguistics. Firstly, he had been the Prussian ambassador in Vienna, where he was involved in shaping the new European order after Napoleon's defeat. Later, at the end of the 1810s, he retired to Tegel, where he spent much of the remainder of his life dealing with extensive linguistic studies.[33]

On the occasion of his death, Humboldt was hailed as a statesman, writer, and educational reformer. But the fact is that he was never a point of reference in nineteenth-century discussions on the university. His fame was to come later. This crucial insight is offered by relatively recent research on Humboldt's legacy that historians such as Mitchell G. Ash, Rüdiger vom Bruch, Sylvia Paletschek and Walter Rüegg have conducted in the last two decades.[34] Of them, the one to argue most strongly for Humboldt's absence from the nineteenth-century debate is Paletschek. With her exhaustive study of the University of Tübingen in the German Empire and during the Weimar Republic as her starting point, she has written about "the invention of Humboldt". On the basis of her research, she claims that Humboldt's programmatic texts remained unknown or even unpublished. In university histories, treatises by Schleiermacher, Fichte and Steffens were invoked that had appeared at the time of the foundation of the University of Berlin. Similarly, other people who were noted in their time, but had fallen into oblivion afterwards, figured largely in accounts of the early nineteenth-century universities. But rarely, if ever, the name of Wilhelm von Humboldt was mentioned.[35]

---

33 Scurla, *Wilhelm von Humboldt* (1976): 608–609.

34 Walter Rüegg, "Der Mythos der Humboldtschen Universität", in: Mathias Krieg and Martin Rose (eds.), *Universitas in theologia – theologia in universitate. Festschrift für Hans Heinrich Schmid* (Zürich: Theologischer Verlag 1997): 155–174; Rüdiger vom Bruch, "A Slow Farewell to Humboldt? Stages in the History of German Universities, 1810–1945", in: Ash (ed.), *German Universities, Past and Future* (1997): 3–27; Sylvia Paletschek, *Die permanente Erfindung einer Tradition. Die Universität Tübingen im Kaiserreich und in der Weimarer Republik* (Stuttgart: Steiner Verlag 2001); Vom Bruch, "Die Gründung der Berliner Universität" (2001); Sylvia Paletschek, "Verbreitete sich ein 'Humboldt'sches Modell' an den deutschen Universitäten im 19. Jahrhundert?", Schwinges (ed.), *Humboldt international* (2001): 75–104; Sylvia Paletschek, "The Invention of Humboldt and the Impact of National Socialism: The German University Idea in the First Half of the Twentieth Century", in: Margit Szöllösi-Janze (ed.), *Science in the Third Reich* (Oxford: Berg 2001): 37–58; Marc Schalenberg, *Humboldt auf Reisen? Die Rezeption des "deutschen Universitätsmodells" in den französischen und britischen Reformdiskursen (1810–1870)* (Veröffentlichungen der Gesellschaft für Universitäts- und Wissenschaftsgeschichte 4) (Basel: Schwabe Verlag 2002).

35 Paletschek, "Verbreitete sich ein 'Humboldt'sches Modell'" (2001): 94–96.

FIGURE 8.1 Friedrich-Wilhelms-Universität *around 1880, by F. Albert Schwartz, in: Harald Brost and Laurenz Demps,* Berlin wird Weltstadt. Photographien von F. Albert Schwartz, Hof-Photograph *220.*
BERLIN: BRANDENBURGISCHES VERLAGS-HAUS 1997

Nor was the University of Berlin considered a beacon in the academic archipelago. The *Friedrich-Wilhelms-Universität*, as it was renamed in 1828, was just one university among many others in the German-speaking region (see figure 8.1). Nothing in its statutes intimated that a new type of university had seen the light of day. Despite the calls for a new order, the faculty hierarchy was unchanged: theology, law, medicine, and finally philosophy. Even with regard to the administrative structure, the forms of examination, and the professors' specialisms, Berlin did not differ markedly from other universities in the German area. Like other new or reformed universities it was an institution financed by the state, even though Humboldt himself had argued for a more traditional financial basis (land and prebends) to ensure that it had a measure of independence from the state.[36]

No more was Berlin a model in nineteenth-century intellectual debates. Works of reference, encyclopaedias and compendiums did not refer to the neo-humanist ideas or Prussian university reforms as turning points. Rather, the modern university's birth was credited to the Enlightenment rationalism of Göttingen and Halle. History books emphasized the widespread death of

36 Paletschek, "Verbreitete sich ein 'Humboldt'sches Modell'," (2001): 79–80.

the universities in the eighteenth century, the dramatic changes in administrative structure at the start of the nineteenth century, the nascent academic liberalism, and the emergence of students' associations around the middle of the nineteenth century. In none of these contexts was the foundation of the University of Berlin acknowledged to have had a shred of paradigmatic significance. It was merely mentioned en passant, often in the same breath as the newly founded universities of Bonn (1818) and Munich (1826).[37]

Throughout the entire nineteenth century, a debate raged about the state of German universities that, judging from the number of published treatises, reached its most intense phase in the 1830s and 1840s, even though much was published on the subject around 1800 and again in the final decades of the nineteenth century as well. It was generally agreed upon that the qualities that set apart Germany's universities were academic freedom and the theoretically strong, scholarly teaching. From time to time, the neo-humanist university tracts of the early nineteenth century were remarked upon, Schleiermacher's publication in particular, but usually the focus of the debate was rather on concrete problems such as examinations, forms of study and the lecturers' working conditions.[38]

Another distinguishing characteristic of the German universities was thought to be the combination of scholarship; unlike in France, all subjects were accommodated in one single university. In itself, this was not particularly new; on the contrary, in a historical perspective it can be seen as a strong element of continuity from the fundamental ideal of the medieval university. In all important respects the old hierarchy survived. Only a limited number of professors argued that the philosophical faculties should have the controlling or uniting role at nineteenth-century German universities. When the universities' mission was first defined, the emphasis was still on the communication of knowledge and the training of lawyers, theologians, and physicians. Only at the very end of the nineteenth century was research brought to the fore. Before then, it was not the majority of professors who argued that the systematic generation of new knowledge was an academic concern of the highest importance.[39]

Consequently, one cannot speak of the presence of either a Berlin model or a Humboldtian model in German academic debate during the nineteenth century. It is true that the University of Berlin was mentioned as a young and dynamic seat of learning, where growth and further expansion was set in relation to the defeat in the Napoleonic Wars and the Prussian capital's

37 Paletschek, "Verbreitete sich ein 'Humboldt'sches Modell'" (2001): 97–98.
38 Paletschek, "Verbreitete sich ein 'Humboldt'sches Modell'" (2001): 98–100.
39 Paletschek, "Verbreitete sich ein 'Humboldt'sches Modell'" (2001): 96–98.

ever-increasing importance, yet it exercised a negligible influence on the German university system's development in the nineteenth century, be it ideologically or institutionally. The ideals that Wilhelm von Humboldt had formulated in 1809 and 1810 neither stimulated a discussion about the universities, nor materialize in their organization on the ground.

Together with gymnasiums, military service and classical music, universities were to be Germany's great export success of the imperial period. In the last decades of the nineteenth century, modern research universities on the German pattern were founded in both Europe and North America.[40] Johns Hopkins University, founded in Baltimore in the 1870s, was to be the first American university with the stated aim of combining an academic education with scholarly research, for example in the shape of a dedicated graduate school. In the same period, a series of young researchers were appointed at other leading American universities, and so contributed to the spread of the new model. But, it should be noted, it was the German university that was the model, which, moreover, was always adapted to indigenous needs and interpretations. Wilhelm von Humboldt was not mentioned.[41]

Rediscovery

When the German Empire was proclaimed in 1871, nineteen universities existed within its borders. In the following half century until the outbreak of the First World War, the number of students quadrupled, but only three new universities were founded: Strasbourg in 1872, Münster in 1902, and Frankfurt am Main in 1914. Strictly speaking, higher education was within the scope of the individual provinces. However, the universities as institutions were seen as a national undertaking and were hotly debated by the pan-German public. Within the borders of the German Empire, academic norms were the same, and both students and professors moved freely between seats of learning. All

40 Schalenberg, *Humboldt auf Reisen?* (2002); Charle, "Patterns" (2004); Edward Shils and John Roberts, "The Diffusion of European Models Outside Europe", in: Rüegg (ed.), *A History of the University in Europe* (2004): 163–230; Jürgen Osterhammel, *Die Verwandlung der Welt: Eine Geschichte des 19. Jahrhunderts* (München: Beck 2009): 1132–1147.

41 Roy Steven Turner, "Humboldt in North America? Reflections on the Research University and Its Historians", in: Schwinges (ed.), *Humboldt international* (2001): 289–312; Roger L. Geiger, *To Advance Knowledge: The Growth of American Research Universities, 1900–1940* (New York: Oxford University Press 1986): 7–9.

this contributed to a sense that Germany's universities made up a coherent, national system.[42]

Even so, it was Prussia that dominated a united Germany, and not just politically, with an overall majority of students studying in a total of eleven universities. Gradually the name of Friedrich Althoff was connected to the leading Prussian university system. Originally a lawyer, in the quarter century from 1882 to 1907, he put his considerable powers and determination to use at the Prussian education ministry, imposing what would come to be called the "System Althoff". With his tough, unorthodox methods, Althoff intervened in recruitment policies and saw to it that new institutions and chairs were founded. His continuous overarching aim was to raise academic standards in Prussia and to defend Germany's cultural standing. Even in his own time, Althoff stoked up strong feelings. It was said that he was the "Bismarck des deutschen Universitätswesens" and Max Weber, among many others, depicted him as a domineering schemer who poisoned the scholarly climate. Recent research has stressed that Althoff must also be seen as an enlightened bureaucrat, who showed no scruples in the service of the nation, safeguarding scholarship in an age of political polarization. Regardless of how one rates Althoff, it cannot be denied that the Prussian seats of learning, and first among them the University of Berlin, became the jewels in the German academic crown in the years around 1900.[43]

The German universities had been held in high esteem since the foundation of the German Empire. During the two subsequent decades, their standing grew further, and by the end of the nineteenth century they were counted amongst the best in the world. Both from elsewhere in Europe and from North America, students visited German universities, attracted by celebrated professors in all academic fields. In the growing rivalry between the European countries, her leading research milieus became a considerable asset for Germany, and for many Germans, especially those from the educated upper middle classes, the

---

42 Konrad H. Jarausch, "Universität und Hochschule", in: Christa Berg (ed.), *Handbuch der deutschen Bildungsgeschichte: 1870–1918: Von der Reichsgründung bis zum Ende des Ersten Weltkriegs* (München: Beck, 1991), vol. 4: 313–344.

43 Bernhard vom Brocke, "Hochschul- und Wissenschaftspolitik in Preussen im deutschen Kaiserreich 1882–1907: Das 'System Althoff'", in: Peter Baumgart (ed.), *Bildungspolitik in Preussen zur Zeit des Kaiserreich* (Stuttgart: Klett-Cotta 1980): 9–118; Bernhard vom Brocke (ed.), *Wissenschaftsgeschichte und Wissenschaftspolitik im Industriezeitalter: Das "System Althoff" in historischer Perspektive* (Hildesheim: Verlag A. Lax 1991) and Peter Josephson, *Den akademiska frihetens gränser: Max Weber, Humboldtmodellen och den värdefria vetenskapen* (Uppsala: Universitet 2005): 169–173.

FIGURE 8.2 *Rainer Christoph Schwinges (ed.),* Humboldt international: Der Export des deutschen Universitätsmodells im 19. und 20. Jahrhundert *(Veröffentlichungen der Gesellschaft für Universitäts- und Wissenschaftsgeschichte 3): cover.*
BASEL: SCHWABE VERLAG 2001

native academic tradition became a source of great national pride. The university was considered the quintessence of German high culture (see figure 8.2).[44]

Despite the gleam of this golden-age, at the turn of the century German universities were simultaneously beset with inner tension. For instance, the number of students rose rapidly, whereas the number of tenured positions experienced only a negligible increase. This was particularly worrying for the young, qualified *Privatdozenten*, who year in, year out were forced to toil in the shadow of their powerful professors. The rise of what amounted to a

44 Rüdiger vom Bruch, *Wissenschaft, Politik und öffentliche Meinung: Gelehrtenpolitik im Wilhelminischen Deutschland (1890–1914)* (Husum: Matthiesen 1980); Anderson, *European Universities from the Enlightenment to 1914* (2004): 151–161.

Dozentproletariat fuelled the generational conflict which had been in existence for some time, and which had only worsened under the hierarchical academic system. The conflicts, however, also ran much deeper than this, and were fundamentally ideological in nature.[45]

During the 1880s and 1890s, the political divisions at the German universities grew. At the time of the February Revolution in 1848, the majority of students and professors had taken a liberal stance, but after unification they rallied in increasing numbers to the nationalist cause and made German's cause their own. A strong state was seen as a guarantee for strong scholarship. The dramatic social upheavals towards the end of the nineteenth century drove them in increasing numbers to the right. Konrad H. Jarausch has characterized this development as "the rise of academic illiberalism". The German nationalism that many scholars professed drew its strength from Prussian, protestant, and conservative sources. Catholics, Jews, and political radicals found it increasingly difficult to sustain a career, and on occasion they were even pushed out of the academic community altogether.[46]

At the same time, competition was growing between the scholarly disciplines. The height of this was the battle over what was meant by the term *Wissenschaft*. In a speech of 1892, the rector of the University of Berlin, Rudolf Virchow, declared that "the dominance of neo-humanism is broken."[47] Virchow had combined scholarship with a career as a liberal politician. For him it was self-evident that the natural sciences would go hand in hand with progressive social development. And Virchow stood not alone in his viewpoint. During the late nineteenth century, a strong expansion of scientific subjects was evident at German universities; not only an increase in quantity but, in more profound terms, an advance for the methodology and approach to knowledge embodied by the "exact" sciences. Keeping pace with Germany's rapid industrialization in the century's final decades, a series of institutes of technology were founded, although they were viewed with mistrust by the established universities and long constituted a separate sphere in higher education. Still, around 1900, applied science had also found a foothold within the large universities. It was not only social reform and economic hopes that were linked to this development. In the

45 Alexander Busch, *Die Geschichte der Privatdozenten: Eine soziologische Studie zur grossbetrieblichen Entwicklung der deutschen Universitäten* (Stuttgart: F. Enke 1959); Anderson, *European Universities from the Enlightenment to 1914* (2004): 153–154.

46 Konrad H. Jarausch, *Students, Society, and Politics in Imperial Germany: The Rise of Academic Illiberalism* (Princeton: University Press 1982); Notker Hammerstein, *Antisemitismus und deutsche Universitäten 1871–1933* (Frankfurt am Main: Campus 1995).

47 Quoted in Anderson, *European Universities from the Enlightenment to 1914* (2004): 154.

run-up to the First World War, and against the background of growing polarization of the European great powers, there were many intellectuals who put their faith in a marriage of *Macht* and *Geist*. "Military strength and science are the two pillars on which Germany's greatness rests, and it is the Prussian state's duty to provide for them both," concluded the influential ecclesiastical historian and science promoter Adolf von Harnack.[48]

But not everyone was equally enraptured by this combination in the years before the First World War. A large number of professors sensed an escalating internal crisis at the universities. Particularly beleaguered were the leading figures of the human sciences, *die Geisteswissenschaften*. The rise of academic specialisms not only cast into doubt their belief in the unity of scholarship, but also posed a threat to the very idea of *Bildung*, which was in many respects their raison d'être. Instead, the scientific world-view was strengthened and the logic of the industrial order threatened to undermine all spiritual values. Scholars of the humanities did not watch these developments in silence.[49]

In *The Decline of the German Mandarins* (1969), Fritz Ringer defended the idea that a crucial shift in opinion among the German professors had taken place around 1890. These "mandarins", chiefly tenured academics in the humanities and social sciences, had long regarded themselves as the protectors of the cultured state and of civilized society, but now their existence was put into question when even the universities were adapted to the realities of the new age. In a mechanized, commercial Germany, shallow and vulgar through and through, as they described it, many of the mandarins felt alienated. They pulled in their horns, limited themselves to fighting turf wars, rejected the politics and social commitment of the day, and girded themselves in cultural pessimism and anti-modernism. During the Weimar Republic, the mandarins embraced the national–conservative reaction that opposed the democratization of society in general and the universities in particular.[50] Ringer's thesis has been influential, but has not been without its critics. Many have questioned whether his characterization of German academics is really representative. Doubts have been raised about his dating of the turning point as being in the 1890s, and there have been suggestions that his interpretation of the fin de siècle academic mood is too coloured by posterity's experience of Nazism.[51]

---

48 Anderson, *European Universities from the Enlightenment to 1914* (2004): 160.

49 Jan Eckel, *Geist der Zeit: Deutsche Geisteswissenschaften seit 1870* (Göttingen: Vandenhoeck & Ruprecht 2008): 23–28.

50 Ringer, *The Decline of the German Mandarins* (1969).

51 James C. Albisetti, "*The Decline of the German Mandarins* After Twenty-Five Years", *History of Education Quarterly* 34 (1994), no. 4: 453–465.

Despite these objections, much of Ringer's general description of the academic climate seems to hold true. The picture of mandarins in retreat fits well with a wider account of Germany in the transition from the nineteenth to the twentieth century. It was a time of expansion and dynamism in many areas; a period of confidence in the future; when ambitious dreams and large hopes were pinned on science. But it was also a time of growing discontent and increasing bewilderment. Which place would the university have in the future? What was the purpose of education? Which spiritual values should be instilled into the nation's mind?

In this turbulent milieu, Humboldt was reborn. In preparing a book on the statesman Wilhelm von Humboldt, the historian Bruno Gebhardt made a hugely significant discovery in the final year of the nineteenth century. In an archive he found the unpublished memorandum *Über die innere und äussere Organisation der höheren wissenschaftlichen Anstalten in Berlin*, which he published in 1903.[52] Suddenly Humboldt's short, incomplete paper of 1809/1810 stood at the centre of the debate on the university at the turn of the century. The text could not only be used to justify fundamental research in general. It was equally useful to sanction the existence of research universities, which had started to be established in Germany only from the beginning of the twentieth century. Humboldt gave this type of knowledge-institution almost century-old credentials. The passage of time had shown that he had been correct.

Somewhat paradoxically Humboldt's rediscovered memorandum gave ammunition to those who wanted to see research institutes kept outside the universities and who favoured, in other words, a separation of knowledge acquisition and knowledge communication. Traditional universities were facing new demands, indeed due to the growth of large-scale research (*Großforschung*) in technology and natural sciences. To maintain Germany's scientific standing in the hardening international climate, in 1911 the *Kaiser-Wilhelm-Gesellschaft zur Förderung der Wissenschaften* was founded. When its first president, Adolf von Harnack, spoke in defence of the new independent research institute, he invoked Humboldt's name. Harnack pointed out that Germany had to thank the Prussian educational reformer for his scientific brilliance and, according to his interpretation, Humboldt asserted in his memorandum that scientific progress required academies, universities, and reasonably independent institutes. Of course not coincidentally, for Harnack, the latter were synonymous with the kind of autonomous research institute, viz. the Kaiser-Wilhelm-Institute that he was in

52 Paletschek, "Verbreitete sich ein 'Humboldt'sches Modell'" (2001): 77.

the process of establishing. It is indeed a general phenomenon that those wishing to create something, try to invoke tradition, as Rüdiger vom Bruch has pointed out. At the turn of the twentieth century Humboldt represented such a tradition; and specifically a tradition one could appeal to if one wanted to create something new in the academic world, without rejecting a supposedly strong legacy.[53]

The success story of the University of Berlin found its proudest expression on the occasion of the celebration of its centenary in 1910. Of particular importance in this regard was the educationalist and philosopher Eduard Spranger. In spirited speeches and in print he cemented the idea that Wilhelm von Humboldt was the author of the modern German university. In *Über das Wesen der Universität*, Spranger collected other neo-humanist documents from the beginning of the nineteenth century and showed their significance in the present. In his view, one could not explain the University of Berlin's flourishing existence without referring to political liberalism, the Prussian reforms and the German nation. Better than anyone else, Humboldt had reconciled academic freedom with interests of the state.[54]

However, it is conspicuous that "the invention of Humboldt" took place precisely in the period when German universities increasingly distanced themselves from the model that had taken shape at the start of the nineteenth century, and which at the end of the same century would become associated with Humboldt's name. The rapidity with which academic specialisms took hold undermined the idea of the unity of science, and the link between research and teaching was weakened as well. At the institutes of technology, but also at an increasing number of universities, research became focussed on practical application. At this stage, Wilhelm von Humboldt as an educational reformer began to take on mythical proportions, in particular in the eyes of Berlin academics. He became a figure from who the various colliding interest groups could draw their energy. For

53 Bernhard vom Brocke, "Die Kaiser-Wilhelm-Gesellschaft im Kaiserreich", in: Rudolf Vierhaus and Bernhard vom Brocke (eds.), *Forschung im Spannungsfeld von Politik und Gesellschaft: Geschichte und Struktur der Kaiser-Wilhelm-/Max-Planck-Gesellschaft* (Stuttgart: Deutsche Verlags-Anstalt 1990): 17–26; Bernhard vom Brocke, "Die Entstehung der deutschen Forschungsuniversität, ihre Blüte und Krise um 1900", in: Schwinges (ed.), *Humboldt international* (2001): 367–401; Vom Bruch, "Die Gründung der Berliner Universität" (2001): 68.

54 Paletschek, "Verbreitete sich ein 'Humboldt'sches Modell'" (2001): 100–102. For Spranger, see Alban Schraut, *Biographische Studien zu Eduard Spranger* (Bad Heilbrunn: Julius Klinkhardt 2007).

Harnack, he became a historical alibi to justify a break with tradition. For the mandarins, with Spranger in the lead, Humboldt's ideas about the university spoke as much to the defence of a threatened order as to a promised vitalization of academic life. Humboldt appealed to all of them. One obvious explanation for these different interpretations, is that his texts on educational philosophy were only rough outlines and so were relatively open to that. Another explanation was that, in an age when German nationalism was on the ascendant, Humboldt as a Prussian official was more appealing as the progenitor of the modern university than, for instance, the theologian Schleiermacher.[55]

It must be underlined, however, that around 1900, it was first and foremost in Berlin that Humboldt was hailed as the father of the German university. In his analysis of university rectors' inaugural addresses (*Rektoratsreden*), Dieter Langewiesche has shown that for a long time it was only in the Empire's capital that Humboldt's name was mentioned on such occasions. The notion that the German university model could be traced back to the foundation of the University of Berlin was championed only there. Langewiesche calls this body of historical writing "die Berliner Ursprungserzählung", and characterizes it as one of the building blocks in the general Prussian–German national mythology. After the fall of the German monarchy, it became increasingly difficult to defend this narrative.[56]

Humboldt During the Interwar Years

The First World War was a disaster in many respects for the development of German universities. Academics from all disciplines had been swept up by war fever and had enthusiastically offered their services to the nation. Both students and professors volunteered for the army, many of whom falling in battle. This obviously had a huge impact upon ordinary academic work, which naturally suffered greatly as result. The battle lines criss-crossed the academic

55 Christoph Markschies, *Was von Humboldt noch zu lernen ist: Aus Anlass des zweihundertjährigen Geburtstags der preussischen Reformuniversität* (Berlin: University Press 2010): 111–127.

56 Dieter Langewiesche, "Die 'Humboldtsche Universität' als nationaler Mythos: Zum Selbstbild der deutschen Universitäten im Kaiserreich und in der Weimarer Republik", *Historische Zeitschrift* 290 (2010), no. 1: 53–91; Dieter Langewiesche, "Humboldt als Leitbild? Die deutsche Universität in den Berliner Rektoratsreden seit dem 19. Jahrhundert", *Jahrbuch für Universitätsgeschichte* 14 (2011): 15–37.

world and the conflict left a deep rift in the international scholarly community.[57] Yet, with regard to the landscape of the German university, the First World War did not result in a complete break with academic tradition. The main institutions remained unaltered. During the period 1919–1939, only two new universities (Hamburg and Cologne) were founded, in both cases immediately after the end of the Great War. The collapse of the German Empire, the ideological schisms of the interwar period, and the Nazi seizure of power were to have lasting consequences for the interpretation of the Humboldtian tradition, however.[58]

During the 1920s, it was noticeable that German scientists had lost status. They had compromised themselves by their involvement in the war, and many of them were now shunned by the international scientific community. Some individual researchers could still enjoy high repute and maintain international contacts, but as a collective the German academics were damaged goods. Moreover, the interwar economic crises eroded the universities' finances, and periodically it became difficult to find the means to buy books, journals, and laboratory equipment. To remedy these financial problems, a number of organizations was founded at the beginning of the 1920s, amongst them the predecessor of the *Deutsche Forschungsgemeinschaft*. Those who set the tone – personalities such as Adolf von Harnack and Fritz Haber – were convinced that Germany could regain lost ground if there was a concentrated effort to invest in technology and science. To an extent they were proved right. German research did indeed recover relatively quickly, but at the cost of a complete and permanent reversal in the academic system: far more research would be conducted outside the universities, and far fewer resources would go to the humanities.[59]

This shift in core values aggravated the existing conflict between *Naturwissenschaften* and *Geisteswissenschaften*. It would, however, be wrong to reduce this to a battle between "the two cultures". One of the arguments in Ringer's work on the German mandarins, after all, is that, somewhat schematically, there were two types of academics: the orthodox, who made up the

57 Notker Hammerstein, "Epilogue: Universities and War in the Twentieth Century", in: Rüegg (ed.), *A History of the University in Europe* (2004): 637–672 and Trude Maurer, *Kollegen – Kommilitonen – Kämpfer: Europäische Universitäten im Ersten Weltkrieg* (Stuttgart: Steiner Verlag 2006).

58 Hartmut Titze, "Hochschulen", in: Dieter Langewiesche and Heinz-Elmar Tenorth (eds.), *Handbuch der deutschen Bildungsgeschichte: 1918–1945. Die Weimarer Republik und die nationalsozialistische Diktatur* (München: Beck 1989): 209–212.

59 Vom Bruch, "A Slow Farewell to Humboldt?" (1997): 19–20.

majority, completely rejected the new form of government and were deeply unsettled by industrialized civilization, while the modernists (among them Weber, Tönnies, Troeltsch, Mannheim and Meinecke) had a more ambivalent attitude to the new democratic state. In the political polarization that marked the 1920s and early 1930s, academic ideals became one element in the ideological struggle, and many of the students rallied to the National Socialist banner. The universities were never bulwarks of democracy.[60]

With the fall of the Prusso-German monarchy, the way in which the history of the University of Berlin was presented altered. Dieter Langewiesche points out that the First World War was a turning point in this respect. Humboldt was gradually freed from the Prussian–national tradition and began to be embodied more and more within the science-organization university model.[61] In the interwar debate on education and scholarship, Humboldt seems to have been evoked in various, partly contradictory contexts, certainly not always by name, and often mainly as an idealization of what the German university ought to be.

One group of mandarins invoked the Humboldtian ideal (if not always Humboldt himself) as a corrective to current tendencies. This group included the circles around Eduard Spranger, Werner Jaeger and other essentially orthodox mandarins for whom the Humboldtian tradition held the promise of a "third Humanism" (following the Renaissance and neo-humanism), a strand in German thought that some of them cleaved to with a kind of restorative nostalgia.[62] Others gave it a very different spin. Carl Heinrich Becker, a leading orientalist and several times Minister of Culture and Education in the Weimar Republic, addressed the German university system in several works published in the interwar period. Best known was his pamphlet *Vom Wesen der deutschen Universität* (1925). Becker's starting point was that the Humboldtian legacy ought to be managed and adapted to the new circumstances of the 1920s immediately. The universities in Germany should remain bastions of pure science, but the task was not to sing their praises, but to develop them in order to keep pace with the age. The universities should have a different, more instrumental function in an age of parliamentarian democracy. To this end, Becker recommended among other reform measures that new chairs of sociology be endowed.[63]

---

60 Ringer, *The Decline of the German Mandarins* (1969); Titze, "Hochschulen" (1989): 216–220.

61 Langewiesche, "Die 'Humboldtsche Universität' als nationaler Mythos" (2010); Langewiesche, "Humboldt als Leitbild?" (2011).

62 The phrase "der dritte Humanismus" was coined by Eduard Spranger.

63 Guido Müller, *Weltpolitische Bildung und akademische Reform. C.H. Beckers Wissenschafts- und Hochschulpolitik 1908–1930* (Köln: Böhlau 1991). Among Becker's works on eduction policy are *Gedanken zur Hochschulreform* (Leipzig: Quelle & Meyer 1919); *Vom Wesen*

The Humboldtian tradition was not solely a source of edification or inspiration. Several commentators have noted that the idea of *Bildung* underwent a transformation between 1800 and 1900, from a forward-looking and utopian ideal to a defensive and ideological one. As a consequence, critics of varying persuasion stepped forward to challenge the governing ideal.[64] The most important and well-known was Friedrich Nietzsche, whose anathema of contemporary education resounded in the interwar period. In several works, primarily from the 1870s, the German iconoclast had lambasted the uniformized science, the utilitarian view of knowledge and the sclerotic, life-denying nature of *Bildung*. Instead, he called for what amounted to an old-fashioned education of free spirits; his vision was one of aesthetic and intellectual exploits by a small elite. This Nietzschean stance can be traced throughout Germany's subsequent history of ideas, including education and educational philosophy. During the interwar period, reforming educationalists, avant-garde artists, and the youth movement all drew strength from Nietzsche in their attempts to overthrow the traditional notion of education. In many cases, however, there was no question of expressly criticizing Humboldt, but rather of kicking life into something that was on its last legs.[65]

Did the Nazi takeover in January 1933 mark the next sharp turn in the history of German universities? For many years, this question was answered with a resounding no. The general feeling was that the vast majority of academics had avoided ideological involvement in this period and had gone into a kind of inner exile for twelve dark years. Academic institutions had largely remained intact. Apart from a couple of National Socialist pet subjects – such as prehistory, folklore, and racial science – the regime had not politicized teaching and research. Most of the students had successfully continued their academic studies, as best as they were able to at this difficult time. The conclusion ran that the universities had been apolitical havens in an ideological age.

Only during the last decades of the twentieth century did this picture start to change. Thanks to a series of critical studies, the established version had to be revised. The one that emerged was far from flattering. It transpired that several of the leading post-war scientists had been happy to serve the regime and

---

der deutschen Universität (Leipzig: Quelle & Meyer 1925); and *Die Pädagogische Akademie im Aufbau unseres nationalen Bildungswesens* (Leipzig: Quelle & Meyer 1926).

64 Fritz K. Ringer, *Fields of Knowledge. French Academic Culture in Comparative Perspective, 1890–1920* (Cambridge: University Press 1992): 95–108.

65 Timo Hoyer, *Nietzsche und die Pädagogik: Werk, Biografie und Rezeption* (Würzburg: Königshausen & Neumann 2002); Christian Niemeyer, *Nietzsche, die Jugend und die Pädagogik: Eine Einführung* (Weinheim: Juventa 2002).

to provide the ideological underpinnings for Nazi policy.[66] In the 2000s, this general scrutiny has been continued in a number of critical studies and biographies, which have benefited from the publication of larger contextualizing works. The broader, synthetic approach has opened the way to a comprehensive reconsideration of the chosen terminology, methodology, and research issues in Nazi Germany. It is striking how many academic subjects seem to have displayed the same, shared norms and mindset, without having been the object of state-enforced regimentation in the name of ideology.[67] With regard to the human sciences, Jan Eckel has pointed to a strong element of continuity, evident from the late nineteenth century to the 1960s. The concepts and theories that were prevalent in the 1880s and 1890s survived intact in all essentials into the early post-war period. Eckel's concluding observation is revealing in this respect: the crucial turning-points of high politics – 1914, 1933, 1945 – do not coincide with those of the history of knowledge. Of course, the tenor of teaching and research was affected by the changing circumstances in society at large, certainly during the era of dictatorship, but the general course of events, which he describes, in effect amounts to knowledge-as-process – one in which questions and concepts are superimposed and so undergo a gradual transformation. Revolutions rarely, if ever, take place in universities.[68]

---

66 Michael Grüttner and John Connelly (eds.), *Zwischen Autonomie und Anpassung. Universitäten in den Diktaturen des 20. Jahrhunderts* (Paderborn: Verlag Ferdinand Schöningh 2003); Michael Grüttner a.o. (eds.), *Gebrochene Wissenschaftskulturen. Universität und Politik im 20. Jahrhundert* (Göttingen: Vandenhoeck & Ruprecht 2010); Notker Hammerstein, "National Socialism and the German Universities", *History of Universities* 18 (2003), no. 1: 170–188; Mitchell G. Ash, "Politicizing 'Normal Science' in Nazi Germany", *H-Net Book Review* (2009), www.h-net.org/reviews/showrev.php?id=25865 (date accessed 01/07/2011). In recent decades many individual universities have been the subjects of monographs. See, for example, Steven P. Remy, *The Heidelberg Myth: The Nazification and Denazification of a German University* (Cambridge: Harvard University Press 2002); Uwe Hossfeld a.o. (eds.), *Kämpferische Wissenschaft: Studien zur Universität Jena im Nationalsozialismus* (Köln: Böhlau 2003); Karen Bayer, Frank Sparing and Wolfgang Woelk (eds.), *Universitäten und Hochschulen im Nationalsozialismus und in der frühen Nachkriegszeit* (Stuttgart: Steiner Verlag 2004); Leo Haupts, *Die Universität zu Köln im Übergang vom Nationalsozialismus zur Bundesrepublik* (Köln: Böhlau 2007) and Thomas Becker (ed.), *Zwischen Diktatur und Neubeginn: Die Universität Bonn im "Dritten Reich" und in der Nachkriegszeit* (Göttingen: V & R Unipress/ Bonn University Press 2008).

67 Hartmut Lehmann and Otto Gerhard Oexle (eds.), *Nationalsozialismus in den Kulturwissenschaften*, 2 vols. (Göttingen: Vandenhoeck & Ruprecht 2004). See also Wolfgang Bialas and Anson Rabinbach (eds.), *Nazi Germany and the Humanities* (Oxford: Oneworld 2007).

68 Eckel, *Geist der Zeit* (2008): 131–138.

These insights are crucial if one is to judge the Humboldtian tradition's standing during the Third Reich. For the Jews and political opponents among the German scientists, the Nazi takeover was a watershed. Altogether, between 15 and 20 per cent of all lecturers were sacked, and many of academic world's brightest stars were forced to leave Germany. However, for the majority, the events of the 1930s only disturbed their self-image to a limited extent. The mandarins continued to believe that they were working in a German university system that traced its origins to Wilhelm von Humboldt. True to their stolid, serious discipline, they continued to be faithful to the leading principles of science.[69] Indeed, after 1933 the Nazis attempted to impose *Gleichschaltung* on some areas of academic life, but on the other side there was never any question of a systematic reconstruction of the universities. Nor is it correct to argue that the Nazis had a coherent science or research policy. Ideologically determined research, with *Rasse* and *völkisch* as its key terms, was the business of special institutes outside the universities, such as the *Reichsinstitut für Geschichte des neuen Deutschlands* and the *Forschungsgemeinschaft deutsches Ahnenerbe*.[70]

This does not mean that the Third Reich lacked ideas about a new kind of university. Adolf Rein and Ernst Krieck, for instance, did present some proposals for a reform of the university system. Both had developed their thinking on the nature of universities well before 1933, but their ideas were adopted by the Nazis who duly promoted their cause in Hitler's Germany. In due course, during the 1930s, both also became university rectors. Rein, a national-conservative historian from Hamburg, was a keen supporter of "the political university", a dream of the regeneration of a paralysed institution that met with approval in academic circles. Krieck, an autodidact and influential inter-war educationalist, had in a similar, but more blatantly racist manner established himself as the exponent of "national-political edification". The two men had a good deal in common. Both wanted to transcend the neo-humanist idea of the university and place knowledge at the service of the Greater German

69 Hammerstein, *Antisemitismus und deutsche Universitäten* (1995); Notker Hammerstein, "Humboldt im Dritten Reich", in: Schwinges (ed.), *Humboldt international* (2001): 469–483; Paletschek, "The Invention of Humboldt and the Impact of National Socialism" (2001).

70 Notker Hammerstein, *Die Deutsche Forschungsgemeinschaft in der Weimarer Republik und im Dritten Reich: Wissenschaftspolitik in Republik und Diktatur 1920–1945* (München: Beck 1999); Paletschek, "The Invention of Humboldt and the Impact of National Socialism" (2001) and Anne C. Nagel, *Hitlers Bildungsreformer: Das Reichsministerium für Wissenschaft, Erziehung und Volksbildung 1934–1945* (Frankfurt am Main: Fischer Taschenbuch Verlag, 2012).

nation. Yet both also believed in the pairing of teaching and research, and rejected narrow academic utilitarianism. Krieck in particular, argued that the German universities should distance themselves from the bourgeoisie, nineteenth-century ideology that was associated with Humboldt's name.[71]

No vision of a new university has been as much discussed as the one Martin Heidegger evoked in his infamous inaugural address as rector of the University of Freiburg in April 1933. The research literature on the philosopher's views on Nazism has been apologetic for a long time. His Nazist ideas were labelled as a short-lived incongruity in his biography, as being an evidence of his political naïveté. In more recent studies, however, the approach has been distinctly more critical, and the relationship between Heidegger's thinking and National Socialist ideology has been thoroughly analysed. Two general conclusions may be drawn: firstly, that he had genuine motives for becoming a university rector, and secondly, that he was not an innocent tool of the ruling powers.[72] During his ten months as rector, Heidegger embraced the *Führerprinzip* and publicly espoused a number of other Nazi doctrines. He thought that he could use his office to work towards a kind of spiritual Nazism, and the reason why he resigned as rector in the spring of 1934 had nothing to do with differences of opinion with the political regime. Instead, it had turned out that his vision of a new kind of university had met with opposition from within the traditional academic system, and Heidegger had been branded as quixotic by some of the professors. Still, his fateful address cannot be reduced to a simple manifesto for a Nazi academy. If one applies the perspective of university history to the issue, it becomes apparent that, at least in part, he was drawing on a long German tradition. Like the Romantics of the early 1800s, Heidegger wanted to bring together all forms of knowledge and experience in an all-embracing scholarship; one that would be embodied in the university and hold philosophy as its cardinal virtue. Politics, and indeed actually all aspects of existence, were to be incorporated in this philosophical knowledge. Thus far, Heidegger was

71 Arnt Goede, *Adolf Rein und die 'Idee der politischen Universität'* (Berlin: Reimer 2008); Gerhard Müller, *Ernst Krieck und die nationalsozialistische Wissenschaftsreform: Motive und Tendenzen einer Wissenschaftslehre und Hochschulreform im Dritten Reich* (Weinheim and Basel: Beltz 1978).

72 There is a steadily growing body of literature on Heidegger and Nazism, with the impact of Victor Farías, *Heidegger et le nazisme* (Lagrasse: Verdier 1987) still being felt. The subject continues to spawn debate. As recently as 2010 there was a prolonged exchange in the *Times Literary Supplement* prompted by a critical review of Emmanuel Faye's book *Heidegger, l'introduction du nazisme dans la philosophie: Autour des séminaires inédits de 1933–1935* (Paris: Albin Michel 2005).

operating within the same idealist framework as, for example, Humboldt, albeit using his own distinctive vocabulary. At the same time, there is no doubt that Heidegger's university existed to serve the providential nation. And therefore, in his opinion, there was no place for "much-extolled 'academic freedom'", as "this freedom was not real, since it was merely negative." Instead, the students were to be bound to the people, the nation's honour, and the German people's spiritual calling.[73]

The examples of Rein, Krieck and Heidegger are revealing examples to prove the seductions of ideology and intellectual responsibility. In the present instance they are probably most interesting as illustrations of the fact that even during the Third Reich the Humboldtian tradition functioned as a significant point of reference. Rein's and Krieck's ideas about the university were very much a product of the discussions of the 1920s. In their essentials they remained true to the established academic model, but at the same time they argued that the university should be subordinated to the demands of the nation. Heidegger for his part had a dream, Romantic in origin, of an institution of paramount knowledge. It did not prevent him from jettisoning all ideas of the university as an autonomous body.

Wilhelm von Humboldt thus seems to have been resurrected time and again during the interwar period, by turns as an auspicious prince of light and the embodiment of a defunct form of education. This was one of the arguments for Peter Moraw to develop his, much-cited, three-phase model of German university history – pre-classical (to circa 1800), classical (from circa 1800 to the 1960s) and post-classical (from the 1960s).[74] Rüdiger vom Bruch, however, has problematized this division. According to vom Bruch, one can argue that the break with the classical tradition came earlier, viz. at the start of the 1930s, providing that one does not go solely by the stability of the institutions, but rather looks to the overarching idea of the university. Vom Bruch clearly emphasizes that in Nazi Germany, the ethos of science was undermined and academic freedom was curtailed.[75] Of course, there are many developments

73 Martin Heidegger, "Die Selbstbehauptung der deutschen Universität", in: Martin Heidegger, *Gesamtausgabe: I. Abteilung: Veröffentliche Schriften 1910–1976* (Frankfurt am Main: Vittorio Klostermann 2000), vol. 16: 107–117.

74 Peter Moraw, "Aspekte und Dimensionen älterer deutscher Universitätsgeschichte", in: Peter Moraw and Volker Press (eds.), *Academia Gissensis: Beiträge zur älteren Giessener Universitätsgeschichte* (Marburg: Elwert 1982). Moraw has since changed his views on the chronology somewhat, see Peter Moraw, "Universitäten, Gelehrte und Gelehrsamkeit in Deutschland vor und um 1800", in: Schwinges (ed.), *Humboldt international* (2001): 19–21.

75 Vom Bruch, "A Slow Farewell to Humboldt?" (1997): 23.

that support his argument, but still the Nazi takeover did not simply result in the complete demise of the Humboldtian tradition. It lived on as a foil; as a corrective to *Realpolitik* – and to *Machtpolitik*.

The Humboldtian Tradition and University History

Against the background of the development of the modern German university, a few general conclusions can be made. For instance, any attempt to reform a cultural or social institution is based on a set of assumptions about that institution's mission. In this respect, the idea of the university cannot be seen as "a free-floating abstraction but a guiding conception, rooted in the experiences, traditions, and life-worlds of individuals", as Björn Wittrock has pointed out. Since the Enlightenment, educational ideals have been pondered and repondered in step with the broader changes in society. However, none of the major university reforms should be seen solely as a response to the process of modernity. Rather, "they occurred because leaders, thinkers, scholars, and scientists continually questioned the basic nature and meaning of higher learning," as Wittrock argues.[76]

By far and away the most important of all these "guiding conceptions" was undeniably the one associated with Wilhelm von Humboldt. In the wider perspective, it becomes apparent that the early twentieth-century Humboldtian renaissance was just one in a long series of discussions about the nature of the modern German university, which have continued with varying intensity ever since. Often it has been an issue of the interpretation of a small number of canonized texts from the turn of the nineteenth century. Rarely these exegetists have been historians. Usually they were scholars of the humanities or social sciences with an interest in university or research politics: from Eduard Spranger, Max Weber and Carl Heinrich Becker, through Karl Jaspers, Réne König and Helmut Schelsky, to Hans-Georg Gadamer, Jürgen Habermas and Jochen Hörisch.

The Humboldtian tradition thus left its imprint on twentieth-century German universities for the very reason that it was always open to reinterpretation. In this way something new emerged, which at the same time drew both its name and legitimacy from what had preceded. What was once depicted as a transhistorical idea of the university became a repository of academic dreams – and the chosen ground for intellectual battles. Humboldt became, to quote Sylvia Paletschek, a universal weapon.[77]

76 Wittrock, "The Modern University: The Three Transformations" (1993): 347.

77 Paletschek, "Verbreitete sich ein 'Humboldt'sches Modell'" (2001): 103.

The research on the Humboldtian tradition that has been pursued in recent decades has put into perspective the idea of the university in modern Germany. It has little to do with commonplace ideas about the meaning of the Humboldtian legacy having changed. Instead, it has mapped out how a notion of the university can strike a chord with the most characteristic moments and grandiose dreams of an epoch. Another important lesson to be drawn from the new research is the intimate – and uninterrupted – connection between research policy and university history. Each serious attempt to transform higher education is dependent on a certain degree of historical understanding; each genuine attempt to write university history is also a contribution to a contemporary discussion.

On a more overarching level, the new research has contributed to the intellectual revitalization of an entire field. For too long, too much university history has been either antiquarian or monumental – not least when academics have celebrated their own grand traditions. However, university history must also be critical and contextualising if it is to have wider relevance and carry real weight. People must on occasion summon up the will to see through the smoke screens of the past and free themselves from the tyranny of simplistic notions. And perhaps this is particularly true of academics.

CHAPTER 9

French Academia in a Prosopographic Perspective

A Collaborative Joint Project

Emmanuelle Picard

The history of French universities remains largely unwritten. The national structure of the teaching corps partly explains this phenomenon, as it has played a part in orienting the historiography towards an examination of the disciplines, as opposed to institutions. Only Parisian academics, the profession's elite, have been the subject of profound studies. The lack of work on the academic corps as a whole hinders our ability to fully understand French academia as an institution. In an effort to remedy this, a prosopographic project dealing with the French academic corps in sciences, humanities, and law between 1800–1940, started in 2011. Like all large-scale prosopographic projects undertaken by a sizeable team of collaborators, it required serious preliminary reflection on the kind of data to be collected in order to foster a genuine social history of the academic profession. This article defines the conceptual framework and practices of the study in progress, thereby endorsing it as an alternative way of writing university history in contrast to the traditional jubilee history.

Seen by the Revolutionaries as agents of the Church, the universities in France were suppressed in 1793. Initiatives led by Napoleon I resurrected them in a reduced form fifteen years later.[1] In contrast to what can be observed in French secondary education during the same period,[2] the university's brief disappearance profoundly transformed the nature of the academic professions in France.[3] Although its full development would actually span several decades, we can consequently date the birth of the modern French university to the First Empire. This slow process, which by the 1880s had transformed the professoriate of the Imperial *facultés* into academics in the modern sense of the term, is a rewarding object of study for those interested in the development of

1 Decree of 17 March 1808 creating the Imperial University.

2 See the work of Dominique Julia and Marie-Madeleine Compère on the continuity between the *collèges* of the Ancien Régime and the *lycées* created by Napoleon. Also Dominique Julia, "La naissance du corps professoral", *Actes de la recherche en sciences sociales* 39 (1981): 71–86.

3 The term 'profession' will be used here as 'professional category' or 'group' to describe a population whose central characteristic is exercising its trade in a strict regulative framework and in institutions that form a coherent whole.

 | DOI 10.1163/9789004265073_010

professions. Yet in contrast to the historiographies of other Western European countries, relatively little research has been done on the history of the French academic corps in the nineteenth and twentieth centuries, Christophe Charle[4] and Victor Karady's work[5] (mostly dealing with the Parisian *facultés*) and several regional monographs aside.[6] Moreover, due to the methodological and evidentiary diversity of publications on the subject, this work does not easily lend itself to comparison and at present a synthetic historical vision of the French academic profession remains out of reach.

Given that French academics constitute a centrally managed corps of civil servants, the lack of an overarching narrative seems somewhat paradoxical. Instead of enjoying the kind of autonomy that characterizes the status of their international counterparts, assent to Parisian authority has long typified the position of French academics. Today, the organizing principles of the academic milieu still draw more on vertical (disciplinary), rather than horizontal (establishment based) consideration.[7] The existence of a rich, centralized source – the individual retirement files of public education personnel at the National Archives in Paris – ought to have made possible, if not a systematic study of the entire population, at least a series of comparative monographs drawing on comparable data.[8] However, the underdeveloped nature of the

---

4 Christophe Charle, *La République des Universitaires, 1870–1914* (Paris: Seuil 1994) and Christophe Charle and Régine Ferré (eds.), *Le personnel de l'enseignement supérieur en France aux XIXe et XXe siècles* (Paris: Editions du CNRS-INRP 1985).

5 To cite just one article: Victor Karady, "L'expansion universitaire et l'évolution des inégalités devant la carrière d'enseignant au début de la IIIe République", *Revue française de sociologie* 14 (1973): 443–470.

6 Among several recent studies: Jean-François Condette, *Les lettrés de la République. Les enseignants de la Faculté des Lettres de Douai puis Lille sous la Troisième République (1870–1940). Dictionnaire biographique* (Lille: Université de Lille 3 2006); Laurent Rollet and Marie-Jeanne Choffel-Mailfert (eds.), *Aux origines d'un pôle scientifique: faculté des sciences et écoles ingénieurs à Nancy du Second Empire aux années 1960* (Nancy: Presses Universitaires de Nancy 2007); Bernard Lachaise, "Les professeurs de la faculté des Lettres de Bordeaux de 1914 à 1968: esquisse de portrait de groupe", in: Jean-François Dunyach and François-Joseph Ruggiu (eds.), *Les Passions d'un historien. Mélanges en l'honneur de Jean-Pierre Poussou* (Paris: Presses de l'Université Paris-Sorbonne 2010): 223–235; see also the online ongoing bibliography of the History of Education Service at www.inrp.fr/she/picard_biblio_etablissements_enseignement_superieur.htm (date accessed 19/06/2014).

7 Christine Musselin, *La longue marche des universités françaises* (Paris: Presses Universitaires de France 2001).

8 The individual retirement files of National Education employees are preserved in collection F17 of the National Archives. Those files of individuals born at least 100 years ago can be freely consulted. Files on those born later require special authorization.

history of universities as a subfield in France and a general undervaluation of comparative work has stifled this sort of effort. While many biographies and monographs deal with small groups within the overall profession (particular laboratories or schools of thought), their narrow focus impedes an understanding of the field as a whole.[9]

These sorts of case studies do not represent the totality of the system and therefore fail to address more general questions about the profession. Additionally, Christophe Charle and others have reminded us that the French centralist tradition, present in the organizational logic of higher education, has reinforced the economic and symbolic place of Paris at the summit of the academic hierarchy. Without disputing this claim, it does not suffice to simply focus on the elite segment of the French academic world, which remains far removed from the average experience by definition. To explore the profession – and not just its privileged members – and its development as a whole, it is precisely the average experience that needs examination. This necessitates a large-scale, long-term study that takes all of its members into account.

Such an undertaking would allow us to tackle historical questions that deal with universities as institutions, notably in terms of the commemorative celebrations that have become increasingly common in the last several years. During more than a century, successive reforms have fragmented the history of French universities, complicating an institutional approach to their individual histories.[10] Against this background, the use of a large-scale analysis of the teaching corps provides us with a useful tool to explore the specific fortunes of different establishments. Through it, dialectics between a national academic corps and particular universities can be brought to light, permitting a study of career paths and choices: individual instructors remaining at one institution throughout their career, or conversely, employing strategies meant to earn them a position in Paris.

Our observations led us to invite the French scholarly community to join in a collaborative effort to assemble a systematic, standardized set of data to encourage the study of the academic profession as a whole in addition to its statutory, disciplinary, local and temporal sub-groups. The first push grew out of an effort led by Calire Lemercier (*Centre national de la recherche scientifique*, CNRS),

9 Emmanuelle Picard, "L'histoire de l'enseignement supérieur français: pour une approche globale", *Histoire de l'éducation* (2009), no. 122: 11–34, histoire-education.revues.org/index1938.html (date accessed 19/06/2014).

10 Emmanuelle Picard, "Recovering the History of the French University", *Studium. Tijdschrift voor Wetenschaps- en Universiteitsgeschiedenis/Revue d'Histoire des Sciences et des Universités* 5 (2012), no. 3: 156–169.

to resume Christophe Charle's line of inquiry in his analysis of the Parisian faculty of sciences. A group of legal historians, organized by Jean-Louis Halpérin (*École normale supérieure*, ENS), Catherine Fillon (*Université Lyon III*) and Frédéric Audren (CNRS), soon joined in. While the Napoleonic (re)foundation of the universities beginning in 1802 may have been the natural starting point for the study, the end-date was less obvious. At this stage, it seemed presumptuous to treat the French university after 1960, following the massive numeric expansion of its staff. Likewise a lack of source material makes the post-War period particularly tricky to tackle. Consequently, 1940 became, less by choice than by circumstance, the end-date for our work. Before discussing the practical methods of our prosopography of the French academic corps, we will present its theoretical framework.

Studying the Construction of a Profession

A major lacuna in existing monographs on the subject stems from their inability to conceive of the academic milieu in all of its dimensions. Often, they choose to study the easily identifiable category of chair-holding professors out of convenience. This category, however, is too narrow to result in a proper investigation of what is in fact a vast and diverse professional group. The tendency to concentrate on individuals with dominant positions in the field, moreover, risks furthering the bias that they represent and constitute a norm.

Yet every scholar working on the history of higher education has come across a wide range of ranks and titles – 'substitute professor', 'adjunct professor', 'lecturer', 'assistant', to name a few – corresponding to a variety of situations that sometimes (but not always) lead to the sought-after title of chaired professor. Although bylaws regulate each of these ranks,[11] these texts have never been compiled in an exhaustive reference document, limiting out knowledge of them, and by extension our general understanding of career paths.[12] Unable to compare the relative positions of the individual or individuals under study (where does an adjunct professor stand compared to a substitute professor, for

11 For the nineteenth century, these texts have been compiled and indexed by the Ministry of Public Education in a collection of seven volumes: Arthur Marais de Beauchamp (ed.), *Recueil des Lois et Réglements sur l'enseignement supérieur, 1789–1915*, 7 vols. (Paris: Delalain Frères 1915); available online in the digital collection of the Bibliothèque nationale de France, at gallica.bnf.fr/ (date accessed 19/06/2014).

12 Today, the key reference remains a precious, yet brief article by Françoise Mayeur, "L'évolution des corps universitaires (1877–1968)", in: Charle and Ferré (eds.), *Le personnel de l'enseignement supérieur en France* (1985): 11–28.

instance), most scholars tracing individual careers find themselves at pains to describe its stages. Nevertheless, a purely administrative account of a given career path remains insufficient. In fact, an over-reliance on regulatory documents ignores the disparity between administrative standards and the reality on the ground, leading to homogenized interpretations.[13] An analysis of a professional group that fails to examine its modes for entry and advancement lacks authority when discussing their implementation.[14] By the same token, an approach that neglects to investigating the degree to which practices respected procedures and norms is an empty theoretical exercise. A study that considers the mutual construction of careers and institutions is thus called for.

Given this, we opted to take a twofold approach. First, we initiated an extensive survey of legislative and regulatory documents to reconstruct the legal framework of ranks and promotion practices from the nineteenth century through the 1960s.[15] This endeavour will result in the gradual publication of an online database with an in-depth discussion of how each position evolved over time.[16] Second, a large-scale prosopographic study will shed light on how careers actually developed in daily practice.

---

13 For example, since 1808, faculty professors must hold a doctoral diploma. However, before the creation of the faculties of arts and sciences under Napoleon, neither the *docteur ès sciences* nor *docteur ès lettres* existed. Initially, the *Grand maître de l'Université* (the Minister of Public Education's original title) conferred the title of doctor on an individual he sought to employ in one of the new faculties, often a high school teacher. This practice continued for a number of decades, even as doctoral programmes grew. It links up to a tradition that goes back until the seventeenth century and constitutes one of the origins of conferring honorary degrees, see Pieter Dhondt, "Pomp and Circumstance at the University. The Origin of the Honorary Degree", *European Review of History* 20 (2013), no. 1: 117–136.

14 Which can be expressed in Bourdieusian terms (see the interpretation of the prosopographic approach by Donald Broady, "French Prosopography: Definition and Suggested Readings", *Poetics* 30 (2002), no. 5–6: 381–385) and can also be found in Andrew Abbott's sociology of professions, for example. See Andrew Abbott and Alexandra Hrycak, "Measuring Resemblance in Sequence Data: An Optimal Matching Analysis of Musicians' Careers", *American Journal of Sociology* 96 (1990), no. 1: 144–185. We cannot effectively think about career paths without reference to the space in which they are inserted and the strategies available within that space. The rules that govern such a space, therefore, need to be examined as closely as possible as do its codes and procedures.

15 The 1960s saw important changes in the administration of universities, linked to the abrupt rise in the employment of teachers and the reexamination of traditional positions that no longer conformed to new professional practices.

16 This preliminary work will be made available gradually in the form of an annotated database, accessible through the History of Education Service. The first of these, on doctoral

The disparity between regulations and reality is all the more problematical given that it arose in tandem with the development of the profession. As the teaching corps expanded, new ranks emerged and regulations governing older ones were altered. Through a continual back and forth between the reality on the ground and the profession's organizing principals, a group gradually structured itself around both standards and internal systems of arbitration and negotiation. Careers became organizational in a sociological sense,[17] and as new positions gradually appeared, they reflected a demand for the corps' numeric growth. Consequently, we need a solid understanding of not only each position, but also the credentials of those who obtained it, and its relative place in the hierarchy.

Prosopography has often focussed on questions of networks, families, and social origins. Yet while it is important to bear these factors in mind, they are not the central focus of our project. Rather than study the social position of individuals, we seek to examine the construction of a profession through its specific modalities and its ties to a broader social, economic, political and professional context. Career trajectories are thus our basic unit of study – in particular, how one enters and advances in the profession – its rhythms and the succession or accumulation of positions. Our aim is to reconstruct a typical career path and its evolution in order to analyze each individual's place in the sphere, making it possible to identify and study exceptional cases and reveal implicit constraints on others: we can thus more easily glean particular strategies that resulted in positions at the Sorbonne, for instance, as well as those that impeded them. In this respect, teaching seems like an especially fruitful focal point, since it reveals how academics interacted with the contemporary elite. In addition to providing a source of information on individuals and their social status, this study will demonstrate how French elites functioned more generally from 1800 to 1940.[18]

Delineating a target population for the study was at once the simplest and most challenging task. Having elected to consider the entire teaching

degrees in arts, can be found online at www.inrp.fr/she/theses/scripts/index.php (date accessed 19/06/2014).

17 The term 'organizational career' refers to a sociological model of a linear progression through a single organization's hierarchy.

18 This project was also presented in the framework of a collective examination of the prosopographic method, see Claire Lemercier and Emmanuelle Picard, "Quelle approche prosopographique?", in: Laurent Rollet and Philippe Nabonnand (eds.), *Les uns et les autres... Biographies et prosopographies en histoire des sciences* (Nancy: Presses universitaires de Lorraine 2012): 605–630. An online version with annexes is available at Hal-SHS, halshs.archives-ouvertes.fr/halshs-00521512/fr/ (date accessed 19/06/2014).

corps – defined as all individuals who regularly taught at *facultés*[19] – our first step entailed constructing a list of their employees. Neither the universities themselves, nor the Ministry of National Education possess such a comprehensive roll. A number of sources had to be consulted: instructor registries at the National Archives and the departmental archives tied to specific *facultés* (often incomplete or chronologically limited), published administrative bulletins, course catalogues and flyers, minutes from faculty meetings, etc. Though not definitive, the current list of individuals selected for this study includes over 3500 instructors at the faculties of arts and sciences and approximately 1500 at the faculty of law.[20] Although the list attempts to represent a population, those who only taught occasionally have not been systematically included. Nonetheless, exhaustiveness does not constitute a precondition for a rigorous study. A sample of nearly 5000 subjects remains statistically relevant and the inability of the survey to include *every* case does not call its central findings into question. Moreover, subjects identified later can be added to the database.

To a certain extent, the sources themselves guided how we defined the population. The group is united through employment at the Public Instruction/National Education administration (in 1932, the Ministry of *Instruction publique* changed its name to *Éducation nationale*) and the archives of this body are thus our largest resource. For two reasons, however, the materials found there do not suffice: first, individuals only briefly affiliated with the university often left no trace in these records; second, we believe that in order to fully understand the process of professionalization we have to consider activities that fell outside of the purview of the National Education administration – and so do not figure in its collection. The existing literature on the French professoriate in the nineteenth century shows that many individuals engaged in multiple professional activities linked to

19 The length of instructors' tenure raises problems. By concentrating on regular instructors, meaning those with relatively stable positions, we neglect part-time lecturers, which results in the risk of eliminating alternative career paths from our study. Despite the fact that during the nineteenth century individuals who did not pursue a full-time academic career taught many courses and often had important intellectual roles, we decided to limit our data collection to instructors who taught for at least a year.

20 The study does not deal with the faculties of medicine for two reasons: firstly, the absence of a team of researchers in the field, and secondly, and more fundamentally, the bylaws governing medical faculties were vastly different from those of the other faculties because of the ties of the former to public hospitals. However, a dictionary of members of the medical faculty in Paris does exist: Françoise Huguet, *Les Professeurs de la faculté de medicine de Paris: Dictionnaire biographique, 1794–1939* (Paris: Editions du CNRS-INRP 1991).

their academic identity.[21] These undertakings require special attention. By studying them, we learn about when and how professional academia distanced itself from related fields (such as industry, journalism, and literature).

Yet, as our sources are serial, numerous and often verbose, we had to justify and limit what we consulted. We had to avoid creating an overly heterogeneous database that would make it difficult to compare individuals. But simultaneously, in order to permit a statistical analysis, we had to find comparable data. Thus, we employed materials from archives concerning academic activity and a more limited set of sources dealing with extra-curricular activities. In the end we opted to orient our research largely towards specific materials (notably, the retirement and staff files at the National Archives[22]), but required their systematic examination. At the same time we took advantage of other documents related to education activities, including the alumni yearbooks of the *grandes écoles* and *aggrégation* registries. Both constitute precious resources that complete the information to be found in the staff files.[23] In order to reconstruct activities that took place outside of regular instruction (in industry, legal affairs, journalism, etc.), we turned to sources on groups where at least one instructor in the survey took part in (philanthropic and/or industrial associations, terms served in the government and/or administration, scholarly or professional societies, etc.). We then drafted a list of those that multiple subjects participated in, which will expand as work progresses.

Once we had framed our general hypothesis and targeted a population and a core set of sources, the technical aspects of the project had to be laid out so that multiple research teams could participate in the work. We created a simple, specific set of instructions in order to limit the potential pitfalls typically associated with a large collaborative effort piloted by a small team. By providing contributors with clear directions on what information to mine on extra-university activities, we structured the inquiry in such a way as to avoid amassing exceedingly disparate and incomparable data. Essentially, we set up a simple method for gathering information that could be adapted as the

---

21 See Charle, *La République des Universitaires* (1994).

22 Despite their a priori systematic character, career files are far from uniform, ranging from volumes to a few sheets of paper.

23 Immediately following the French Revolution, an examination for high school teachers was created, which today functions as an unofficial, though generally necessary, step in an academic career. See Emmanuelle Picard and Marte Mangset, "La communauté des historiens académiques français à l'aube du XXIe siècle", in: Christophe Granger (ed.), *À quoi pensent les historiens. Science et insouciance de l'histoire au XXIe siècle* (Paris: Autrement 2013): 31–47.

project progressed. Next, given that our aim was to make the data (and not simply the study's findings) as widely available as possible, publishing it online seemed self-evident, so that others were allowed to make use of our findings. Lyon's LAHRHA history laboratory offered to integrate our work in its more general historical database: SyMoGIH (*Système Modulaire de Gestion de l'Information Historique*) created in 2012.[24] To get the project off the ground as quickly as possible, we provisionally used a non-dynamic data entry form that allowed for regular updates (Excel spreadsheets). Finally, as the scale of the project required the assistance of a large number of scholars, we established clear guidelines for research, selection and entry to maintain a high degree of consistency and ensure accurate quantitative processing. Consistency is a central problem in the design and creation of any database and therefore, a number of methodological handbooks deal with the subject.[25]

The Questionnaire: Rubrics and Overall Philosophy

Having examined the scholarly foundations of the project, we can now address some of the practical aspects mentioned above. The initial data entry uses Excel spreadsheets.[26] The spreadsheets, containing a dozen columns, are divided into five rough categories: identity, education, teaching employment, non-teaching employment and networks/memberships.

- The 'identity' rubric contains basic information of the individual from personal records, including name or names (given name, pseudonym, stage name, etc.),[27] date and place of birth, nationality, social background, marriage status and descendants. Moreover, subjects are assigned a unique

24 See larhra.ish-lyon.cnrs.fr/Pole_Methodes/SyMoGIH_fr.php (date accessed 19/06/2014).

25 An example of this can be found at the History of Education Service's website. On instructors at the Paris faculty of sciences, see www.inrp.fr/she/dictionnaire_faculte_sciences_paris_dossier_complet.htm (date accessed 19/06/2014).

26 The presentation format of this file does not authorize reproduction in this chapter. Any scholar who wishes to consult it, may request it by email (emmanuelle.picard@ens-lyon.fr).

27 This dimension is particularly important in regard to academic careers, given that names have a value themselves both in validating statements and assigning them to an author; see David Pontille, "La signature scientifique. Authentification et valeur marchande", *Actes de la recherche en sciences sociales* 141–142 (2002): 72–78. The legal name may differ from a professional or penname used by an academic to sign publications. Therefore this method ensures that all documents will be linked to the individual in question, regardless of the name cited in a particular source.

identification number that is associated with each item connected to them. Identification numbers, more reliable than proper names that may change over time, make detecting links between individuals easier.

- The 'education' rubric begins when possible with their secondary education, but more often with the higher education that they received, specifying institution(s) attended and diplomas or titles earned (see table 9.1).
- 'Teaching career' distinguishes between teaching in and outside of the *facultés*. It contains all activities and positions held, with precise notes on rank, area(s) taught, and duration of engagement (see table 9.2).
- Due to the suppositions made in the study, 'the individual's non-teaching professional career' is the most difficult to document. It is constructed, like the preceding rubrics, with information on the rank, period, and place of employment.
- The 'networks/memberships' rubric surveys all extra-professional activity: scholarly journals, professional associations, political/trade union and religious activities.[28]

TABLE 9.1 *Fields in the 'education' rubric.*

Doctorate, discipline	Source language
Doctorate, year received	Digital
Doctorate, dissertation title	Source language
Doctorate, number of pages in dissertation	Digital
Doctorate, committee	Free text that indicates the identification numbers of committee members present in the database
Doctorate, honorary (if any)	Source language
Dissertation chair, name	Source language
Dissertation chair, identification	Digital
Doctorate, where defended	Source language
Remarks	

28 In this way we hope to circumvent the tendency that can be found in the 'professional' prosopographic approach to overlook a population's activity outside of their central field. For instance, in his study of nineteenth-century Bologna academics, François Gasnault remarks on his inability to take into account the "innumerable connections that unite [the group] to other institutional networks"; see François Gasnault, "Le milieu universitaire à Bologne au XIXe siècle. Les aléas de l'enquête documentaire prosopographique", *Mélanges de l'Ecole française de Rome. Moyen-Age, Temps modernes* 100 (1988), no. 1: 155–173.

TABLE 9.2 *Fields in the 'teaching career' rubric.*

Starting date of episode 1	Day-month-year
End date of episode 1	Day-month-year
Post title for episode 1	Source language
Instructor title for episode 1	Pull-down menu
Identification number of post of attachment	Digital
If substitute, name of instructor being substituted	Source language
Identification number	Digital
Remarks	

A major problem in creating a prosopographic database for quantitative data processing rapidly became apparent: for at least part of the data to be consistent and systematic it was not enough to direct researchers towards particular sources, a universal approach had to be used for each subject. This left us with a decision: either restrain data collection to a largely serial list of pre-defined sources (potentially significantly limiting the richness of the survey), or use a dual-format to construct the survey, by supplementing data from a required list of sources with other materials. A missing entry for a given variable thus indicates that despite being sought out, information could not be found. In addition to the obligatory fields, contributors to the database can freely introduce additional material in the 'remarks' rubric (see the tables above). By doing so, we ensured that a significant part of the data comes from a systematic search of the target population without losing important complementary information. To round out the system, we decided not to code or pre-categorize the data, asking participants to enter information in the source language (i.e. in full text). This step was intended to prevent anachronistic or ahistorical readings, such as attributing a given rank to the social value it holds today, when during the nineteenth century, for instance, it had a far different status. This precaution seemed especially important for a database that tracks a population over a 150-year span that saw profound changes in professional identity. Our approach also illuminates structural transformations in which the subjects' careers evolved (for example, the development of chairs and disciplines). By taking the operative effects of filiation and appropriation into account, instead of imposing anachronistic twentieth-century categories, it becomes possible to create a genealogy of the organization of academic knowledge.[29]

29 For an example of such critiques, as well as a proposal for overcoming them, see Steven Shapin and Arnold Thackray, "Prosopography as a Research Tool in History of Science: The British Scientific Community, 1700–1900", *History of Science* 12 (1974): 1–28.

Similar factors led us to employ the notion of 'episode' as our data entry mode. The prosopographic approach risks flattening out specific trajectories; by reconstructing an individual's sequential accumulation of positions and credentials (such as membership in several groups or the accrual of recognition), it becomes difficult to recognize the effects and significance of simultaneity. However, in order to analyze a trajectory, we have to be able to show the cumulative or competitive effects that may have impacted it.[30] Identifying when an individual entered a specialist association or received a distinction reveals both how this came about and the ramifications of new credentials. Since a number of synchronic variables often shape academic careers, the concept of 'multi-positioning' through various forms of association and engagement, lends itself to an analysis of how subjects capitalized on their credentials. We can then think about how individuals inserted themselves in networks and furthered their academic careers through (power) positions. Thus the database needed a format where we could reconstruct entire careers as completely as possible, including the accumulation of teaching posts, administrative functions, and work in other spheres. Understood as performing a specific function in a particular place over a period of time, the use of 'episodes' permits the emergence of a detailed reconstruction. Therefore, rather than relying on a catch all rubric like 'teaching experience', each episodic entry contains not only chronological information, but also qualitative content. This underlines a trajectory's synchronic (the existence of simultaneous episodes) and diachronic (career development considerations) dimensions.[31] This practice applies to every rubric from life events such as a second marriage (thus a second episode in the matrimonial rubric) to diplomas (in the educational rubric). Bearing in mind the totality of position-functions over a lifetime, we can view individuals as the products of a complex series of changing relations over time, offsetting the tendency to anachronistically reduce a complex career to a prestigious post occupied late in a career.

This data should allow for the following types of quantitative processing:

---

30 See Luc Boltanski, "L'espace positionnel: multiplicité des positions institutionnelles et habitus de classe", *Revue française de sociologie* 14 (1973), no. 1: 3–26. While this article demonstrates how an approach that encompasses multiple positions can be used to measure an individual's social surface, it does not take time into account, in other words the transformations or sequence of surfaces.

31 On the significance of using smaller units than individuals to structure information, see the methodological work underway at the LAHRHA, in particular Sylvain Brunier and Nicolas Krautberger, "Les trajectoires archivées des experts de la modernisation rurale alpine (XIXe–XXe siècles)", *Temporalités* 11 (2010), temporalites.revues.org/index1251.html (date accessed 19/06/2014).

- factorial analysis of correspondences (possibly with automatic classification) to uncover what contributes to diversity and the existence of disparate profiles in a group, and to statistically identify individuals that typify specific profiles, permitting a rational, targeted return to biography from prosopography;[32]
- multivariate regression and event-based historical analysis to evaluate the impact of various factors on an individual's 'success' (defined according to criteria specific to the group under investigation);[33]
- network analysis to understand how various types of relationships (familial, economic, etc.) and interactions (citation, collaboration, etc.) structured a group and contributed to the creation of favourable positions and divisions in it;[34]
- sequential analysis, which can produce typologies based on the sequence of posts held, places of residence or any other temporal series as opposed to more static 'profiles'.[35]

Conclusion: A Step-by-Step Project

The extent of the survey undertaken – nearly 5000 subjects in all – entailed a collective effort and led us to conceive of a simultaneously rigorous and broad questionnaire to accommodate jurists and physicians alike. Our decision to use an enlargeable online database from the outset increased the project's flexibility and facilitated the long-term recruitment of collaborators. It also gave us the chance to demonstrate the system's effectiveness and reliability by testing

32 See for example Frédéric Lebaron, "La dénégation du pouvoir. Le champ des économistes français au milieu des années 1990", *Actes de la recherche en sciences sociales* (1997), no. 119: 3–26 and Björn-Olav Dozo, "Données biographiques et données relationnelles", *ConTEXTES* 3 (2008), contextes.revues.org/index1933.html (date accessed 19/06/2014).

33 See for example Jérôme Krop, Claire Lemercier and Pierre Schermutzki, "Relations sociales et désignation d'une génération de directeurs d'écoles dans le champ de l'enseignement primaire de la Seine, 1870–1914", *Revue d'histoire moderne et contemporaine* 57 (2010), no. 2: 79–114.

34 See for example Douglas R. White and H. Gilman McCann, "Cities and Fights: Material Entailment Analysis of the Eighteenth-Century Chemical Revolution", in: Barry Wellman and Steven Berkowitz, *Social Structures: A Network Approach* (Cambridge: University Press 1988): 380–399, eclectic.ss.uci.edu/-drwhite/pub/Chemical.pdf (date accessed 19/06/2014).

35 See for example Fabien Accominotti, "Creativity from Interaction: Artistic Movements and the Creativity careers of Modern Painters", *Poetics* 37 (2009): 267–294.

it on a sample of the target population: the Paris faculty of sciences. Limiting ourselves to a specific rank would have prevented an analysis of professional dynamics and the structuring of the field. Consistency, along with the opportunity to fill in a gap in the current literature, motivates our choice as Christophe Charle's collection of dictionaries on the Parisian faculties lacked a volume on this institution during the nineteenth century.[36]

Created after the establishment of the Imperial University on 17 March 1808, the Paris faculty of sciences gradually developed over the next decade. Initially its activities were limited to public lectures and overseeing the distribution of university diplomas. Over the course of the nineteenth century, however, it slowly became a leading institution for both teaching and research. Its growth, already noteworthy under the Second Empire, was particularly spectacular during the early years of the Third Republic. Between 1900 and 1939, the Paris faculty of sciences was a centre of French scientific life. Its faculty included well-known professors (including Augustin Louis Cauchy, Marie Curie and Jean Perrin), a large number of French and foreign students attended courses, and the staff of its research and teaching laboratories trained countless graduate students and participated in major scientific debates. A number of more or less well-researched biographies and analyses have been written on leading personalities at the faculty and their place in the scientific fields. By contrast, other long-forgotten instructors have left little trace behind. The earlier remarks on the development of the academic community and the trajectories and interactions within it force us to think about the teaching staff as a whole and to attempt to compile consistent and, more importantly, comparable data on each of them.

The first difficulty that we faced was to limit our undertaking. As noted above, no single source would have allowed us to survey all individuals who occupied a teaching position for at least a year. By comparing various lists – appointment notices, faculty registers and the like – we identified 318 individuals for the period 1808–1914. We then offered all interested scholars an opportunity to participate in the project by 'adopting' instructors. This approach gave historians and historians of science, whether specialists of a discipline or of a particular figure, responsibility for their area of expertise. To ensure that the serial data would be systematically entered, we began by

36 Christophe Charle and Eva Telkès, *Les Professeurs du Collège de France, 1901–1939. Dictionnaire biographique* (Paris: Editions du CNRS-INRP 1988); Christophe Charle and Eva Telkès, *Les Professeurs de la faculté des sciences de Paris, 1901–1939. Dictionnaire biographique* (Paris: Editions du CNRS-INRP 1989) and Christophe Charle, *Les Professeurs de la faculté des lettres de Paris, 1809–1939. Dictionnaire biographique*, 2 vols. (Paris: Editions du CNRS-INRP 1985–1986).

examining a number of lists (former students of the *École normale supérieure* and the *École polytechnique*, members of the Academy of Sciences and the Academy of Medicine, the *Morts pour la France* database, etc.) and communicated our results and directions on how to locate individual files at the National Archives to those willing to 'adopt' instructors at the Paris faculty of sciences. Scholars involved also received the entry form with detailed instructions on how to fill it out.

This first step is underway. Presently, at the behest of Jean-Louis Halpérin (ENS) and under the direction of Catherine Fillon, a group of legal historians have begun a prosopographic study of instructors at France's faculty of law between 1802 and 1950.[37] Early on, the team agreed to collaborate with the 'Paris faculty of sciences' project. This combined effort was natural and necessary, since the project aims not to understand how a single subsection operates, but the academic world as a whole. Moreover, the project likewise incorporated 'local' initiatives, in particular a study of the Nancy faculty of sciences.[38]

Together this extensive collaboration will result in a cohesive, online, open-access database that will promote exchange among scholars working on socio-historical questions. Although developed for French academia, the prosopographic questionnaire ought to prove sufficiently flexible and robust for use in other national contexts. As the nineteenth century was a crucial period in the emergence of the modern academic profession in Western Europe and the United States, it constitutes a promising basis for international comparison.

Since 2012, the network Heloise (European Workshop on Historical Academic Databases) has promoted scholarly exchange on the social history of European universities from the Middle Ages to the present. It has likewise worked towards creating tools and technical solutions that promote further efficient collaboration in producing, treating, and making digital data and archives available. Its second workshop took place in Bologna in June 2013 and resulted in an academic blog: heloise.hypotheses.org/ (date accessed 19/06/2014). The discussion focussed on how to facilitate a simultaneous search of multiple online databases. This should provide a useful tool for comparison between the European countries implicated in the project and consequently further the history of

37 The presentation is available at www.droit.ens.fr/-Histoire-des-professeurs-et-des-.html (date accessed 19/06/2014).

38 See Laurent Rollet, "Peut-on faire l'histoire des pôles scientifiques?", *Histoire de l'éducation* (2009), no. 122: 91–113, histoire-education.revuews.org/index1946.html (date accessed 19/06/2014).

European universities in the long-term. In result, the project as a whole offers another way of writing university history, in contrast and/or in addition to the traditional jubilee history. Instead of creating a false sense of unity by writing an artificial commemorative history of different institutions that have little more in common than their place of business,[39] the large-scale analysis of the teaching corps in this project, provides us with a useful tool to explore the specific fortunes of different establishments. A comparative approach is therefore essential, both within France and between French and other European institutions.

39 See the example of institutions like the University of Avignon that was created long before the Revolution by the Pope in 1303 and disappeared in 1793, like all the other universities in France. The university lay dormant until 1963 when it became a satellite of the University of Aix-en-Provence, only acquiring the status of a full university in 1984. The University of Avignon inherited no more than a memory of its pre-Revolutionary namesake. Nonetheless, efforts have been made to transcend the century and a half long gap between the two institutions, as in the recent work Brigitte Bénézet (ed.), *L'Université d'Avignon: naissance et renaissance, 1303–2003* (Arles: Actes Sud 2003).

CHAPTER 10

University History as Part of the History of Education

Pieter Dhondt

Even up to the period when the research imperative gradually took hold at Western universities, they were first and foremost considered educational institutions, as is illustrated in the first section of this article by way of the debate on the function of the university in Belgium around 1880. However, despite this clearly educational background, historical studies on the university have never been regarded as being part of the history of education, although this sub-discipline has managed to establish itself rather strongly, mainly through its special place within teacher training. The second section discusses briefly the institutionalisation of history of education as a discipline, and looks at how one of the main concerns in the field, i.e. presentism, can at the same time be used as an opportunity to defend the field, also with regard to university history writing. Following up on this, the main ambition of the chapter, particularly in its third and last section, is to explain what university history as a sub-discipline could gain by connecting itself more closely to methodologies and concepts used in the field of history of education, such as classroom history, the integration of a larger variety of (visual and material) sources, conceptual frameworks like segmentation, grammar of schooling, demythologization, educationalization and the pedagogical paradox. The last notion in particular points to the relativity of the often overblown rhetoric with respect to the field of education, also applicable to the level of universities.

Research, teaching and service to society is the generally accepted triad of tasks within the modern university, and is generally accepted in that order in terms of appearance and importance, at least in the eyes of the university authorities. However, had the same considerations been taken into account at the end of the nineteenth century, at a time when the Western research university had gradually taken hold,[1] the triad would have run very differently: vocational training, scientific schooling and liberal education. By way of the debate within Belgian universities around 1880, it will be shown in the first section of this article that even though research had gradually established itself within

1 William Clark, *Academic Charisma and the Origins of the Research University* (Chicago: University of Chicago Press 2006).

 | DOI 10.1163/9789004265073_011

universities, up until the end of the nineteenth century it was first and foremost considered an educational institution. In a long-term perspective, the research imperative has only been introduced fairly recently, as is confirmed in several chapters within this volume. In spite of this clearly educational background, historical studies on the university have never been regarded as being part of the history of education, although this sub-discipline has managed to establish itself rather strongly, mainly through its special place within teacher training. The second section will discuss briefly the institutionalisation of history of education as a discipline, and look how one of the main concerns in the field, i.e. presentism, can at the same time be used as an opportunity to defend the field, also with regard to university history writing. Following upon this, the main ambition of the chapter, particularly in its third and last section, is to explain what university history as a sub-discipline could gain by connecting itself more closely to the methodologies and concepts used in the field of history of education.

The University as an Educational Institution

Just as their colleagues from many other European countries and the United States had done, Belgian nineteenth-century professors looked with increasing interest and admiration to German universities, certainly after the German unification of 1870–1871.[2] And just as their foreign colleagues did, they were searching for the backdrop on which this German success was based. The Liège professor Louis Jean Trasenster took up a somewhat provocative standpoint concerning the explanation of German scientific supremacy. He attributed it to causes other than just a number of specific features of the German university system, such as better preparation of the students, the greater freedom of students and professors, or the stronger research culture. In an anonymous pamphlet of 1873, he considered the second half of the nineteenth century, "one of these great periods of time in which the axis of the civilised world has been moved". Due to increasing ultramontanism and due to the difficult relationship between religion and science, Latin countries blocked each form of scientific progress, according to Trasenster. The German scientific world on the other hand was characterised by the greatest intellectual freedom. "And this spirit of intellectual independence has penetrated even into the universities of the Catholic parts of Germany," he stated.[3]

---

2 See pp. 200–202.

3 Louis Jean Trasenster, *De l'enseignement supérieur en Belgique* (Liège: Desoer 1873): 12–13.

The origin of German supremacy was, according to Trasenster, the fact that after the crushing defeat in Jena in 1806, Prussia "had understood that it had to regenerate itself, and the salvation had to come from education. It is principally the education at the universities that transformed and saved Germany."[4] His colleague at the faculty of medicine in Liège, Léon Fredericq, agreed with this view a few years later. In his eulogy on the physiological training at the university of Berlin, he repeated the legendary, prophetic words of Frederick William III, on the occasion of the foundation of the university of Berlin in 1810: "The state must replace with spiritual strength that which it has physically lost."[5] Sixty years later the prophecy would be fulfilled. In 1871, German science had defeated the French, just as the German armies had defeated the French, this was how the professors in Liège and many of their colleagues from home and abroad analysed the situation.

Trasenster was certainly not alone in his analysis of the problem. Many of those involved agreed that the lack of freedom was the main cause for the arrears in Belgian university education. However, not everyone found a connection with the increasing orientation towards authoritarian Rome. Most of them searched for an explanation in the typically Belgian contrast between state and free education, on the one hand the state universities of Liège and Ghent and on the other hand the Catholic University of Leuven and the liberal free-thinking University of Brussels. When in 1878 the more progressive Pope Leo XIII succeeded the extremely conservative Pius IX, Trasenster moderated his opinion too.

Moreover, the majority of professors and politicians thought that Belgium should not entirely renounce its (Catholic) Latin and French past. On the contrary, the country should make better use of its central position between allegedly practical France and philosophical Germany, by combining French honesty and accuracy with German inventiveness. The Belgian universities had to search for a happy medium between the (French) applied-practical and the (German) fundamental-scientific approach of education. As one of the few, Trasenster also added England to the picture:

> Belgium, placed in the midst of the three great nations and the three great civilisations of Western Europe, has to cling on to cultivate the qualities that distinguish and unite these different traditions, in a kind of eclecticism: the clearness, the precision and the talent to express oneself of the French; the firm reason and the energetic initiative of the English;

4 Trasenster, *De l'enseignement supérieur en Belgique* (1873): 12.

5 Léon Fredericq, "L'enseignement de la physiologie à l'université de Berlin", *Revue de Belgique* 3 (1881), no. 38: 119. See p. 188.

> the persevering self-abnegation with which the Germans devote themselves to scientific work and the most abstract studies.[6]

However, gradually the demand for a more research-oriented approach in education and the demand for more freedom, which was closely connected to this, became so overwhelming that the Belgian government had to meet these requests at least to some extent. Auguste Beernaert, who had received a travel scholarship to visit the universities of Paris, Berlin and Heidelberg and who later became a famous politician, was convinced that Belgium should take up a middle position with regard to academic freedom. On the one hand, the freedom to establish independent universities was greater than in Germany, but on the other hand, government interference in the practical organisation of education was at least as excessive as during the French occupation of the Southern Netherlands.[7]

Many professors concluded that the French institutions of higher education were gradually enjoying more internal freedom. Belgium could not stay behind any longer, it was stated. As a result of these pleas, lecturers received the permission to organise optional subjects more easily from the 1880s, although the academic and/or political authorities retained a great deal of supervision, since each initiative had to be approved by the government, the Brussels council of administration or the episcopate (with regard to the Catholic University of Leuven). Moreover, for the students, real freedom of choice was still absolutely out of the question. Indeed, they could take optional subjects, but as before, these were not included in the examination programme. Like many others, François Collard, professor in Leuven, asserted that the Belgian students were not (yet) able to deal with the German *Lernfreiheit*.

To a certain extent, as the demand for more freedom was met, the need to advance science and research as an assignment of university education also increased. But still, it was claimed that Belgian universities should not go too far in this direction. As Belgium had to make use of its central position to find a middle course between France, Germany and England with regard to the approach of education, it also had to combine the assignments of the universities that stood at the centre in each of these countries: an excellent vocational

6 Louis Jean Trasenster, "Discours prononcé le 12 octobre 1880, dans la séance d'ouverture solennelle des cours de l'université de Liège (Du rôle de l'enseignement supérieur et des améliorations et compléments qu'il réclame en Belgique)", in: Jean Joseph Thonissen (ed.), *Situation de l'enseignement supérieur donné aux frais de l'état. Rapport triennal. Années 1880, 1881 et 1882* (Bruxelles: Gobbaerts 1886): 104.

7 Auguste Beernaert, "Rapport sur l'état de l'enseignement du droit en France et en Allemagne (1851–1852)", *Annales des Universités de Belgique* 10–11 (1851–1852): 1727–1728.

training as in France, a profound scientific schooling as in Germany and a solid liberal education as in England, according to Trasenster, who voiced the opinion of many of his colleagues.

Although, around 1880, there existed almost unanimity concerning these three tasks of university education as a basic principle, much dissension arose about the practical application of it. Many professors and politicians warned about extreme specialization as in Germany, which would include the risk of educating unworldly scholarly recluses. The University of Leuven in particular stressed its importance as the upholder of general education. It suggested reserving the scientific *cours pratiques* for the best students, because the others, who aimed at a vocational degree, would not be interested in these courses anyway. Stimulated by some prominent professors at the faculty of medicine, the University of Brussels pushed forward the idea of vocational training as the central task. The universities should concentrate on that in which they were really strong, viz. the training of practitioners. In that case, the advancement of science and research should be taken care of in a separate institute. This suggestion was inspired by material and pragmatic considerations also. In result of the transfer of scientific education to a separate institution, less pressure would be put on the already limited budget of the university.[8]

According to the professors, the university was thus in the first place an educational institution, certainly up to the beginning of the twentieth century, ideally combining vocational training, scientific schooling and liberal education. Nevertheless their research activities have received much more attention in the commemoration publications and in university history writing in general than their teaching activities. The differentiation of the disciplines and the increasing professionalisation of university graduates in subjects like history, sociology or natural sciences have been popular topics of study in this regard.[9] In confirmation of this trend and precisely in order to meet this deficiency, a recent project at the KU Leuven (with the title "A relationship of reciprocity? The changing position of teaching and research within the nineteenth-century history professorship") described its ambitions as follows:

8 For more details, see Pieter Dhondt, *Un double compromis. Enjeux et débats relatifs à l'enseignement universitaire en Belgique au XIXe siècle* (Gent: Academia Press, Ginkgo series 2011).

9 By way of example, see Rudolf Stichweh, *Zur Entstehung des modernen Systems wissenschaftlicher Disziplinen: Physik in Deutschland, 1740–1890* (Frankfurt am Main: Suhrkamp Verlag KG 1984), Andrew Abbott, *The System of Professions: An Essay on the Division of Expert Labor* (Chicago: University Press 1988) and Gabriele Lingelbach, *Klio macht Karriere. Die Institutionalisierung der Geschichtswissenschaft in Frankreich und den USA in der zweiten Hälfte des 19. Jahrhunderts* (Göttingen: Vandenhoeck & Ruprecht 2003).

> Studies in the research activities of nineteenth-century history professors have greatly multiplied in recent years. It has been shown that those activities were stimulated by the creation of a solid job market for historians, the formation of academic communities, the rise of archives, and the many professional and friendly relations between historians world-wide. However, there has been relatively little analysis of the part history professors played as teachers. [...] By focussing on the 'practices' of teaching, I want to sketch a much more accurate picture of the nineteenth-century history professorship than hitherto has been drawn in the literature.[10]

Even at German universities, where, for instance in Göttingen and Halle, already from the eighteenth century it was gradually required that professors should not only teach but also conduct research, the research imperative was only properly adapted at the very end of the nineteenth century, as Östling has proven in his chapter.[11] As in Brussels, again financial concerns were at the basis of this development, indeed in this case starting from an opposite argumentation. One of the primary reasons why the Prussian and other German states encouraged and promoted research in their universities was, simply economic. Officials and ministries of education and culture appreciated the difficulties in maintaining a dual system of institutions: academies for research and universities for training.[12] Moreover, also the incorporation of research into the university as an educational institution did not always go so well. As with any innovation, institutionalising research and research-oriented pedagogy has to be learned through trial and error and adapted to local conditions, often provoking criticism. However, it is precisely this process which has been of little interest to university historians, even though they could have made use of the expertise built up through the sub-discipline of history of education.

History of Education as a Sub-Discipline

Especially until the 1990s, collaboration between historians of education and university historians had been minimal, mainly due to the affiliation of the

10 See www.kuleuven.be/research/researchdatabase/project/3H11/3H110337.htm (date accessed 22/06/2014).

11 See p. 199.

12 Charles E. McClelland, *State, Society and University in Germany 1700–1914* (Cambridge: University Press 1980), R. Steven Turner, "The growth of professorial research in Prussia, 1818 to 1848 – Causes and context", *Historical Studies in the Physical Sciences* 3 (1971): 137–182 and Robert Marc Friedman, *Integration and Visibility: Historiographic Challenges to University History* (Oslo: University of Oslo 2000): 9.

former to the faculties of education, in combination with the universal downgrading of pedagogic research, which itself stems from the pragmatic orientation of educational sciences and the unevenness of research quality.[13] Historians of education were reproached for having received an inadequate training in history and of being tempted by a kind of applied history of education, in clear contrast to the (acclaimed) more fundamental, philosophical approach within university history. In accordance with this implicit assumption of a scientific hierarchy, both groups only rarely published in each others domain, they did not meet at the same conferences and only occasionally were historians used as opponents in educational dissertations and vice versa.

Many of these reproaches were prejudices rather than real deficits within the history of education and, cynically enough, similar reproaches could have been made towards the jubilee tradition within university history writing. Nevertheless, the gap between both sub-disciplines remains in existence up to today, though indeed in recent years somewhat less pronounced. In result, the institutionalisation of both fields occurred completely separately. Despite the allegedly inferior position of the history of education, its institutionalisation process[14] started earlier and has been much more successful compared to the field of university history: a larger number of earlier established societies, publishing their own journals,[15] and being united in the International Standing Conference in the History of Education (ISCHE) that

13 Jukka Rantala, "History of education threatened by extinction in Finnish educational sciences", in: Jesper Eckhardt Larsen (ed.), *Knowledge, Politics and the History of Education* (Studies on Education 2) (Berlin: LIT Verlag 2012): 55.

14 See the research project "History of Education: Mapping the Discipline" that, according to its own website "will focus on the emblematic traits that characterize any discipline: its institutional foundation (institutes, departments, posts), communication networks (associations, scientific events, means for publication), the structures of socialization and education of the new generation (curriculum, diploma, doctoral theses) and the ongoing renewal of knowledge produced by the discipline (research, epistemological foundation, research methods). It aims to describe the recent evolution of History of Education in order to make it more visible and, in knowing it and in reflecting upon it, to reinforce its foundation and legitimacy." See kartografy.wordpress.com/about/, offering also a systematic bibliography concerning the historiography of the discipline (date accessed 22/06/2014).

15 E.g. *History of Education Quarterly* (USA, from 1961); *Paedagogica Historica* (Belgium, from 1961); *Journal of the Australian and New Zealand History of Education Society*, continued (from 1983) as *History of Education Review* (Australia/New Zealand, from 1972); *History of Education* (UK, from 1972); *Histoire de l'Education* (France, from 1978) and *Historical Studies in Education* (Canada, from 1984). See William Richardson, "British Historiography of Education in International Context at the Turn of the Century, 1996–2006", *History of*

organises yearly conferences with up to 250 participants from all over the world.[16] The striking difference between the institutionalisation of both fields led to the famous university historian Sheldon Rothblatt's complaint that the history of universities never managed to establish itself as an independent sub-discipline, but instead has remained "something of an institutional orphan".[17]

However, history of education owes its strong position (though also its inferior reputation in the eyes of the historians) largely to its place as an obligatory subject within teacher training, introduced mostly at the end of the nineteenth century and integral in many European countries still. A course in the 'history of education' was seen as fulfilling perfectly the ideal triple task of the university as an educational institution, as sketched above: scientific schooling, by combining history with heuristics, methodology and encyclopedics in the German tradition;[18] liberal education, by providing the future teachers with an historical background on their country and the professional field in which they would end up; and (somewhat intermingled of course) vocational training, by using history for practical educational purposes, such as drawing inspiration and motivation from the examples of the past, and by providing ideas and conceptions to be used as building blocks for a contemporary theory of education.[19]

Over a long period the discipline has been blamed particularly for the last specific approach. Historians of education were accused of 'presentism', of presenting the past as an inevitable progression towards ever greater liberty and enlightenment, culminating in the introduction of compulsory education and the gradual democratization of education at all levels. They would have acted as servants of a particular pedagogical practice, theory, or idea instead of

Education 36 (2007), no. 4–5: 573 and 590. Compare this with the societies and journals within university history, see p. 7.

16 Christoph Lüth, "Entwicklung, Stand und Perspektive der internationalen Historischen Pädagogik am Beginn des 21. Jahrhunderts – am Beispiel der International Standing Conference for the History of Education (ISCHE)", in: Petra Götte and Wolfgang Gippert (eds.), *Historische Pädagogik am Beginn des 21. Jahrhunderts. Bilanzen und Perspektiven: Christa Berg zum 60. Geburtstag* (Essen: Klartext Verlag 2000): 81–107.

17 Sheldon Rothblatt, "The writing of university history at the end of another century", *Oxford Review of Education* 23 (1997), no. 2: 152.

18 Pieter Dhondt, "Die Berliner Universität als Idealbild für belgische Studenten bis zum Ende des Ersten Weltkriegs", *Historische Zeitschrift* 296 (2013), no. 3: 647.

19 Marc Depaepe, "The Ten Commandments of Good Practices in History of Education Research", *Zeitschrift für pädagogische Historiographie* 16 (2010), no. 1: 31.

adopting the critical attitude that can be expected from 'proper' historians.[20] And although currently the history of education is generally standing for historic research of – and not for – pedagogy, the tradition still leaves its traces. For instance, as late as 1988, Richard Aldrich from the renowned Institute of Education at the University of London set out on a course of research projects designed explicitly to reconnect with a professional audience around the organising theme of "historical perspectives on current educational issues", offering in that way a kind of applied educational history in order to counter the isolation of historians of education among educationalists.[21]

Currently one of the main opponents against this orientation is Marc Depaepe, professor in the history of education at the KU Leuven and former ISCHE president. In his opinion "the added value of [...] history of education consists of nothing more than pure wisdom – there are no concrete lessons to be drawn from the past." Simultaneously he does recognise that "presentism is not a methodological 'sin' but rather an unavoidable condition of research in the history of education," as runs another of his "ten commandments of good practices in history of education research".[22] Indeed, history writing always involves a presentistic and perspectivistic viewpoint of which the historian has to be aware and the pitfalls of which should be avoided as much as possible. In that way presentism is no longer regarded as the enemy of good history, but accepted as integral to the condition of writing history. Besides, in consequence of postmodernism, pedagogy has lost its normative character (at least partly), resulting in there being no concrete lessons to be drawn from it, neither in the past nor in the present.

This does not mean that the struggle for the history of education has been won,[23] because notably from the teaching side there is still the ever returning request of a clear link to the present. In reaction to Marc Depaepe's 'decalogue' the Japanese professor Toshiko Ito points to the occupational duality of historians of education. In their research they should draw a sharp line between the past and the present. Yet in their teaching historians also should consciously

20 Marc Depaepe, "How should the history of education be written? Some reflections about the nature of the discipline from the perspective of the reception of our work", *Studies in philosophy and education* 23 (2004), no. 5: 335.

21 Richard Aldrich, "Historical Perspectives on Current Educational Issues", *Bulletin of the History of Education Society* 41 (1988): 9–10 and Richard Aldrich, *The End of History and the Beginning of Education* (London: Insitute of Education 1997): 5. See Richardson, "British Historiography of Education in International Context" (2007): 587.

22 Depaepe, "The Ten Commandments" (2010): 34 and 32.

23 Gary McCulloch, *The Struggle for the History of Education* (Foundations and Futures of Education) (New York: Routledge 2011).

relate the present to the past. They should be able to perceive the past as the seed from which the present sprang. According to her own experience, the students are even explicitly encouraged in this expectation, if only by the course evaluation form, which asks them if the contents of a course addresses present problems and if it has contributed towards improving teaching practices.[24]

Notwithstanding, it is certainly possible to make a link to the present in one's teaching without lapsing automatically into the presentistic approach as defined above. Even more, the pressure on historians of education to create such a link in order to make their students – being future teachers – enthusiastic for a subject that they did not choose and of which they do not see the relevance, has resulted in many historians of education reflecting on the value of their subject. In that way, one of the main concerns in the field, i.e. presentism, became at the same time an opportunity to defend the field. The question of why we should teach history of education developed into a well-researched topic in itself. Some of the regularly returning arguments are: to provide impetus to deal with generally complex, sometimes paradoxical or ironic, and often problematic outcomes of the past; to develop greater analytical acuity when dealing with today's educational problems; to communicate a critical awareness of how education has been used and misused in the past; to show the relativity of the often overblown rhetoric with respect to the field of education to counter the neo-liberal trends of management and efficiency thinking; to cultivate the utility of the non-utilitarian; or to transcend the short-sightedness of our own time.[25]

As having been a lecturer in the history of education myself (at Ghent University from 2009 to 2012), I have been obliged to do the same exercise. In addition to previous ambitions, my personal aim has always been to question the casualness of existing pedagogical beliefs and practices through a historical reflection, and to stir up the same kind of astonishment as when being introduced into ideals and realities from abroad. Indeed, things do not have to be as they are. Of course, I am completely indebted in this regard to the proverbial opening sentence of Leslie Poles Hartley's novel *The Go-Between* (London: Hamish Hamilton 1953): "The past is a foreign country: they do things differently there." However, for the students this often proved to be a much less evident lesson than it may seem. On the basis of exactly the same arguments,

---

24 Toshiko Ito, "Historians and the Present: on Marc Depaepe's Decalogue", *Zeitschrift für pädagogische Historiographie* 16 (2010), no. 1: 43–44.

25 E.g. Kadriya Salimova and Erwin V. Johanningmeier (eds.), *Why should we teach history of education?* (Moscow: International academy of self-improvement 1993) and Larsen (ed.), *Knowledge, Politics and the History of Education* (2012).

a historical course could be included in all other university programmes, in particular in those fields which also have to get rid of the threat of presentism, such as medicine.[26] And when a course in the history of education should be part of teacher training, why then should a course in university history not be made compulsory for future professors?; if only in order to make them acquainted with the different tasks and the universal background of the university as an educational institution.

Some university historians even want to go a step further and argue for an explicitly present-focussed approach, also in their research, inspired or not by the example of history of education. Having studied the long-term historical background of the Bologna process, according to Anne Rohstock "such a present-focused history of higher education could provide us with a behind-the-scenes look at today's political reform programs and social debates, resurrect forgotten historical connections, and with a ground in historical understanding, point to possible future trends in the development of higher education."[27] This plea is actually less spectacular than it may seem on first sight. As late as 1996, the respected Dutch university historian Willem Frijhoff confirmed that all good university history writing has a prospective merit thanks to its broad, problematising approach.[28] And also the current boom in university history writing is certainly, to some extent, intended as an (indeed mostly very implicit or even unconscious) answer to the current crisis of the university, to offer a counterweight in the context of neo-liberalism, increasing commodification and the almost exclusive focus on research.[29]

---

26 E.g. Jacalyn Duffin, "A Hippocratic Triangle: History, Clinician-Historians, and Future Doctors", in: Frank Huisman and John Harley Warner (eds.), *Locating Medical History. The Stories and Their Meanings* (Baltimore/London: The Johns Hopkins University Press 2006): 432–449.

27 Anne Rohstock, "The History of Higher Education – Some conceptual remarks on the Future of a Research Field", in: Daniel Tröhler and Ragnhild Barbu (eds.), *The Future of Education Research: Education Systems in Historical, Cultural, and Sociological Perspectives* (Rotterdam: Sense Publishers 2011): 99.

28 Willem Th. M. Frijhoff, "Hoezo universiteitsgeschiedenis? Indrukken en stellingen", *Ex Tempore. Historisch Tijdschrift KU Nijmegen* 15 (1996), no. 43: 21.

29 Christian Krijnen, Chris Lorenz and Joachim Umlauf (eds.), *Wahrheit oder Gewinn? Über die Ökonomisierung von Universität und Wissenschaft* (Würzburg: Königshausen & Neumann 2011). E.g. the current interest in the historical background of the Humboldtian tradition (see the chapter of Johan Östling) or the choice of subjects in consequence of the geographical widening in university history in general (see pp. 6–7).

What University History can Learn from the History of Education

Of crucial importance in increasing the desired impact of this kind of research is to be sure that policy makers use the historical information correctly. Especially in the first part of this book, many examples have dealt with how the authors of jubilee publications have been mobilised for the purpose of the political, cultural or scientific goal of the university authorities or political administration in organising or controlling the jubilee. And as mentioned in the introduction, many jubilee histories had their origin in offering an answer to a particular crisis of the university, with all its consequences.[30] Without moving towards applied university history writing, history of education could probably still provide some useful tips, having much more experience in this regard.[31] Perhaps one of the most evident lessons, though at the same time most difficult to avoid, is that policy makers will continue to use historical perspectives, but they do so primarily to advance their own agendas.[32] The rhetorical (but further inane) use of the name of Humboldt by rectors and administrators on the occasion of university celebrations is a perfect illustration of this trend.[33]

Secondly, the so-called new cultural history of education from the 1980s has proven convincingly that history does not have to sell something: neither itself, by playing up to those who commission the research; nor its subject, i.e. education, by creating a non-existent linear progress story. By paying attention to the macro-, meso- and micro-level of pedagogical practice, the new cultural history of education has shown how schools modify reforms and make them their own. By starting from the school reality on the pedagogical shop-floor in the classroom, the American historians Larry Cuban, David Tyack and William Tobin took back the history of education to the core of its business as it were:

---

30 See pp. 2–3, 9 and 11.

31 See Maris A. Vinovskis, *History and Educational Policymaking* (Yale: University Press 1999).

32 Rubin Donato and Marvin Lazerson, "New Directions in American Educational History: Problems and Prospects", *Educational Researcher* 29 (2000), no. 8: 9.

33 Anne Rohstock, "Some things never change: The invention of Humboldt in Western higher education systems", in: Pauli Siljander, Ari Kivelä and Ari Sutinen (eds.), *Theories of Bildung and Growth. Connections and Controversies Between Continental Educational Thinking and American Pragmatism* (Rotterdam: Sense Publishers 2012): 165–182 and Pieter Dhondt, "'Humboldt' in Belgium. Rhetoric on the German university model", in: Thomas Karlsohn, Peter Josephson and Johan Östling (eds.), *The Humboldtian Tradition: Origins and Legacies* (Scientific and Learned Cultures and Their Institutions 12) (Leiden: Brill 2014): 97–110.

to research how people in the past really brought up, educated and taught their children. Their studies draw attention to the modernisation resistance of the field and point to the large degree of continuity and stability concerning (teacher-oriented) educational habits.[34] Both aspects are part of the paradox of the modernisation of education: it is not the reforms initiated from above which change the school, but the school bending them to its will. The school transforms them according to its own grammar of schooling, being the formal rules of the specific school and class. The result is an extremely slow evolution of teaching habits and methods in which modernisation myths can fit in.[35]

Without a doubt a similar approach could also be beneficial for university history writing, combining and confronting the higher educational policy at the national and international level, with the policy of the local institutions and finally the application of these measures by the faculties, departments and eventually by the individual professors in the lecture room. As Friedman indicated in his programmatic essay on writing the history of the University of Oslo, the introduction of the research paradigm is quite often uncritically considered a sign of progress, but what was lost in consequence of this development? Indeed, women were gradually admitted into the universities from the second half of the nineteenth century, but what type of male culture emerged in the ever larger laboratory installation opposing the emancipation of women? How did the rise of small armies of doctoral students, research assistants, and other non-teaching staff create shock waves within universities disrupting structural and cultural equilibrium?[36] In a methodological vein, university history would have much to learn from this kind of classroom history.

Another important merit of the new cultural history of education in general, and classroom history in particular, is the use of a large variety of source material, from all kinds of written documents (textbooks, manuals, the pedagogical press, preparations of lessons, inspection reports, school regulations, examinations, school newspapers and magazines of ex-pupils researching the subjective perception of the school mentality), through visual material such as

34 Larry Cuban, *How teachers taught: Constancy and change in American classrooms, 1880–1990* (New York: Teachers College Press 1993); David Tyack and William Tobin, "The grammar of schooling: Why it has been so hard to change?", *American Educational Research Journal* 31 (1994): 435–479 and David Tyack and Larry Cuban, *Tinkering toward Utopia. A century of public school reform* (Harvard: University Press 1995).

35 Marc Depaepe, "Geen ambacht zonder werktuigen. Reflecties over de conceptuele omgang met het pedagogische verleden", in: Angelo Van Gorp, Pieter Dhondt, Frank Simon and Marc Depaepe (eds.), *Pedagogische historiografie. Een socio-culturele lezing van de geschiedenis van opvoeding en onderwijs* (Leuven: Acco 2011): 41–42.

36 Friedman, *Integration and Visibility* (2000): 22 and 17.

pictures and films,[37] to the school 'material' culture (including wall charts, didactical material, school furniture, architecture, etc.).[38] The equivalent of these material sources at the level of higher education – academic heritage – has become a popular subject of study as well, however in almost completely separate circles. Whereas school museums mostly actively engage historians of education and are present at the same meetings, in contrast directors of university museums meet at occasions and network separately from traditional university history, such as Universeum – European Academic Heritage Network, the Association of Academic Museums and Galleries, University Museums and Collections or the Dutch Academic Heritage Foundation.[39] An increasing number of attempts to bridge this gap can be noticed, e.g. the approach at Ghent University in the build-up to its jubilee in 2017[40] and conferences like "Universität der Dinge. Akademisches Sammeln in der Diskussion" (Göttingen, 4–6 October 2012) or "Positioning Academic Heritage. Challenges for Universities, museums and society in the 21st Century" (Ghent, 18–20 November 213). However, also in this respect university history might gain from following the example of history of education more explicitly, by increasing the use of visual and material sources in particular.

Despite, or precisely thanks to the use of a large variety of very concrete source material, history of education as a discipline is often criticised for fancying a high degree of descriptive facts. Therefore, Depaepe included in his "commandments of good practices" the call for improving the interpretative qualities of the research by developing theoretical and conceptual frameworks from within the history of education in order to introduce more structure into the chaos of the educational past.[41] Yet again, already only the existence of a

37 E.g. Paul Warmington, Angelo Van Gorp and Ian Grosvernor (eds.), "Education in motion: uses of documentary film in educational research" – special issue of *Paedagogica Historica. International Journal of the History of Education* 47 (2011), no. 4.

38 E.g. Martin Lawn and Ian Grosvernor, *Materialities of Schooling: design, technology, objects, routines* (Oxford: Symposium Books 2005).

39 See www.aamg-us.org/, publicus.culture.hu-berlin.de/umac/, www.universeum.it/ and www.academischerfgoed.nl/; all websites accessed 11/06/2014.

40 Patrick Monsieur, Jean Bourgeois a.o., "Academic heritage at Ghent University: the field's outlook on the future", *Mededelingen van het Museum voor de Geschiedenis van de Wetenschappen Universiteit Gent* 4(2011): 3–17 and Fien Danniau, Ruben Mantels and Christophe Verbruggen, "Towards a Renewed University History: UGentMemorie and the Merits of Public History, Academic Heritage and Digital History in Commemorating the University", *Studium. Tijdschrift voor Wetenschaps- en Universiteitsgeschiedenis/Revue d'Histoire des Sciences et des Universités* 5 (2012), no. 3: 179–192.

41 Depaepe, "The Ten Commandments of Good Practices" (2010): 33.

journal like *Bildungsgeschichte. International Journal for the Historiography of Education*, encouraging reflection on "the historiography question as to how histories of education can and should be written",[42] proves that the situation can not be that bad after all. In fact, history of education has indeed developed some theoretical, methodological and historiographical concepts that could definitely be advantageous for university history.[43]

In addition to those mentioned above, one of the older examples is the model of systematisation and segmentation that Detlef Müller, Fritz Ringer and Brian Simon provided with respect to secondary education,[44] and that is certainly also applicable to the level of higher and university education, e.g. with regard to the formation of academic disciplines. More recently, Depaepe himself and others developed the concept of educationalization, referring to the increased and continuously increasing efforts in the fields of upbringing, education and schooling in the modern society, and the pedagogical paradox inextricably bound up with this notion.[45] Education increasingly revealed a dynamic of its own and the increase of educational opportunities did not necessarily provide increased opportunities for empowerment or a greater emancipation of the individual. Therefore, demythologizing and deconstructing stories about history seems to be a never-ending task in the history of education. In consequence, the story of progress concerning education has to be forsaken. More education does not automatically result in a better educated, more emancipated, more democratic and wealthier society. Since policy makers, school directors and university rectors are not particularly keen on hearing such a message, "historical researchers are not the best speakers at jubilees," to quote Depaepe again.[46]

---

42 Rebekka Horlacher, Jürgen Oelkers and Daniel Tröhler, "Editorial", *Bildungsgeschichte. International Journal for the Historiography of Education* 1 (2011), no. 1: 7. See also Marie-Madeleine Compère, *L'histoire de l'éducation en Europe. Essai comparatif sur la façon dont elle s'écrit* (Berlin/Paris: Peter Lang/ Institut national de récherche pédagogique 1995) and Gary McCulloch and Ruth Watts (eds.), "Theory, methodology and the history of education" – special issue of *History of Education* 32 (2003), no. 2.

43 Of course, the exercise could be done in the opposite way also, looking for theoretical, methodological and historiographical concepts that are developed within university history and that could be advantageous for the history of education (e.g. concerning the transnational transfer of knowlegde), but that is beyond the scope of this article.

44 Detlef Müller, Fritz Ringer and Brian Simon (eds.), *The Rise of the Modern Educational System. Structural Change and Social Reproduction, 1870–1920* (Cambridge: University Press 1987).

45 Marc Depaepe, *Between educationalization and appropriation. Selected writings on the history of modern educational systems* (Leuven: University Press 2012).

46 Depaepe, "The Ten Commandments of Good Practices" (2010): 33.

If only for that reason, history of education is useful and necessary as a barrier against the hypertrophy of one-sided, utilitarian-designed educational research.[47] Or, as Konrad Jarausch already put it in 1986, the new history of education "was [and had to be] explicitly critical of the celebratory Whig tradition and looked at educational institutions and processes in a self-consciously radical way."[48] That twenty-five years later, Depaepe, the editors of *Bildungsgeschichte* and many other key figures in the field, still repeat the same appeal proves that regarding history of education as a part of historical research is less evident than it may seem upon first sight. Just as university history still has both a fruitful and limiting relationship with jubilee history,[49] history of education is equally ambiguously related to educational sciences. History of education should be respected as a reflective, irrelevant historical science, in contrast to the ruling adage of empirical and actionable educational research.

A closer co-operation between historians of education and university historians could in that way be advantageous for both. History of education could gain some prestige through a stronger connection with historical science, yet following on from the idea of the university as an educational institution, in this article we have made the exercise primarily in the other direction. Absolutely without making a plea for the transformation of university history into an applied science, it would probably do no harm to reflect sometimes a bit more explicitly on how research results could be used (or misused) by policy makers. Just as Jesper Eckhardt Larsen has encouraged his fellow historians of education, opening a discussion on the raison d'être of university history "can be a first step in a productive dialogue between researchers and research politicians on the future politics of knowledge within this specific field."[50] Following up on the finding that jubilee histories have functioned at least to some extent as an answer to a particular university crisis, in the current context, by paying a historically legitimate degree of attention to the educational function of the university, upcoming university jubilees might contribute to a broader, more open-minded view on the institution (in contrast to the dominant neo-liberal, commodified, exclusively research oriented approach).

---

47 Marc Depaepe, "After the ten commandments... the sermon? Comments on David Labaree's research recommendations", *Bildungsgeschichte. International Journal for the Historiography of Education* 2 (2012), no. 1: 89.

48 Konrad H. Jarausch, "The Old 'New History of Education': A German Reconsideration", *History of Education Quarterly* 26 (1986): 225.

49 See p. 10.

50 Larsen (ed.), *Knowledge, Politics and the History of Education* (2012): 11.

Of course, regarding university history as part of the history of education is just one of many approaches of writing history, which have to exist side by side and ideally even in combination. Yet in this article we have tried to show to what extent some theoretical, methodological and historiographical concepts used in the field of history of education might be of advantage to university history as well: classroom history, the integration of a larger variety of (visual and material) sources, conceptual frameworks like segmentation, grammar of schooling, demythologization, educationalization and the pedagogical paradox. The last points to the relativity of the often overblown rhetoric with respect to the field of education. A final revealing example with regard to this is also the inflationary shift in meaning that notions as university, high school and academy underwent during the previous centuries. A long-term approach with more attention for the relationships between different levels of education is therefore called upon to make clear that the development of education, be it on the level of early childhood education, primary, secondary, higher or university education, is not a simple story of progress.

Index

www.ingramcontent.com/pod-product-compliance
Lightning Source LLC
LaVergne TN
LVHW010641300926
849161LV00005B/1013